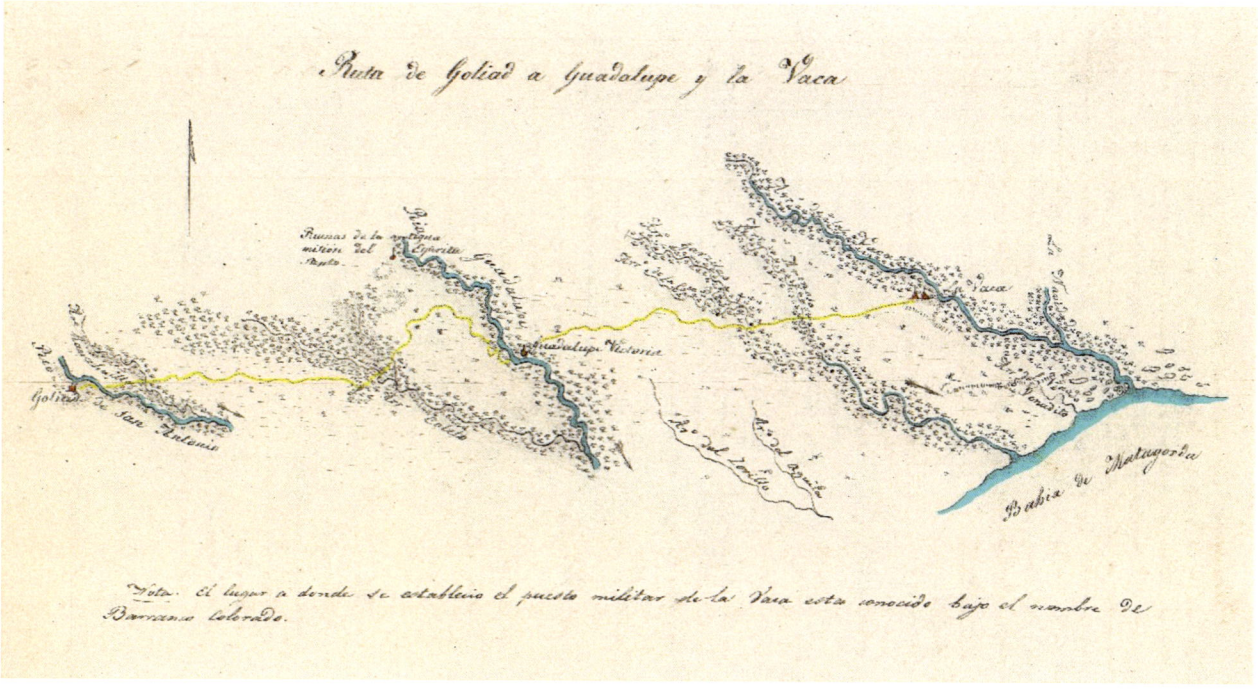

Cover art is from the papers of Jean Louis Berlandier (1803-1851), a Swiss-Mexican naturalist and artist. As a young man, he came to Mexico to make botanical collections, and then was hired to join the *Comisión de Límites*, a boundary-confirming and scientific expedition to Texas that would take about one and a half years from late 1827 to early 1829, led by General Manuel de Mier y Terán. Also on the expedition was Rafael Chowell, its geologist and mineralogist. These other two men will later play an outsized role in the story contained within this book. Berlandier would later return to Texas (Goliad area) in the period of 1834, when the above drawing is thought to have been made (also used below in Figure 31). Today, it is archived at the Beinecke Rare Book and Manuscript Library of Yale University, in a bound volume with others of his map drawings. A prolific note taker and cartographer during this period, his journal was later translated and published in 1980 as "Journey to Mexico during the Years 1826 to 1834". After his travels to Texas, he settled at Matamoros, where he married and also became a physician. He died there by drowning in 1851.

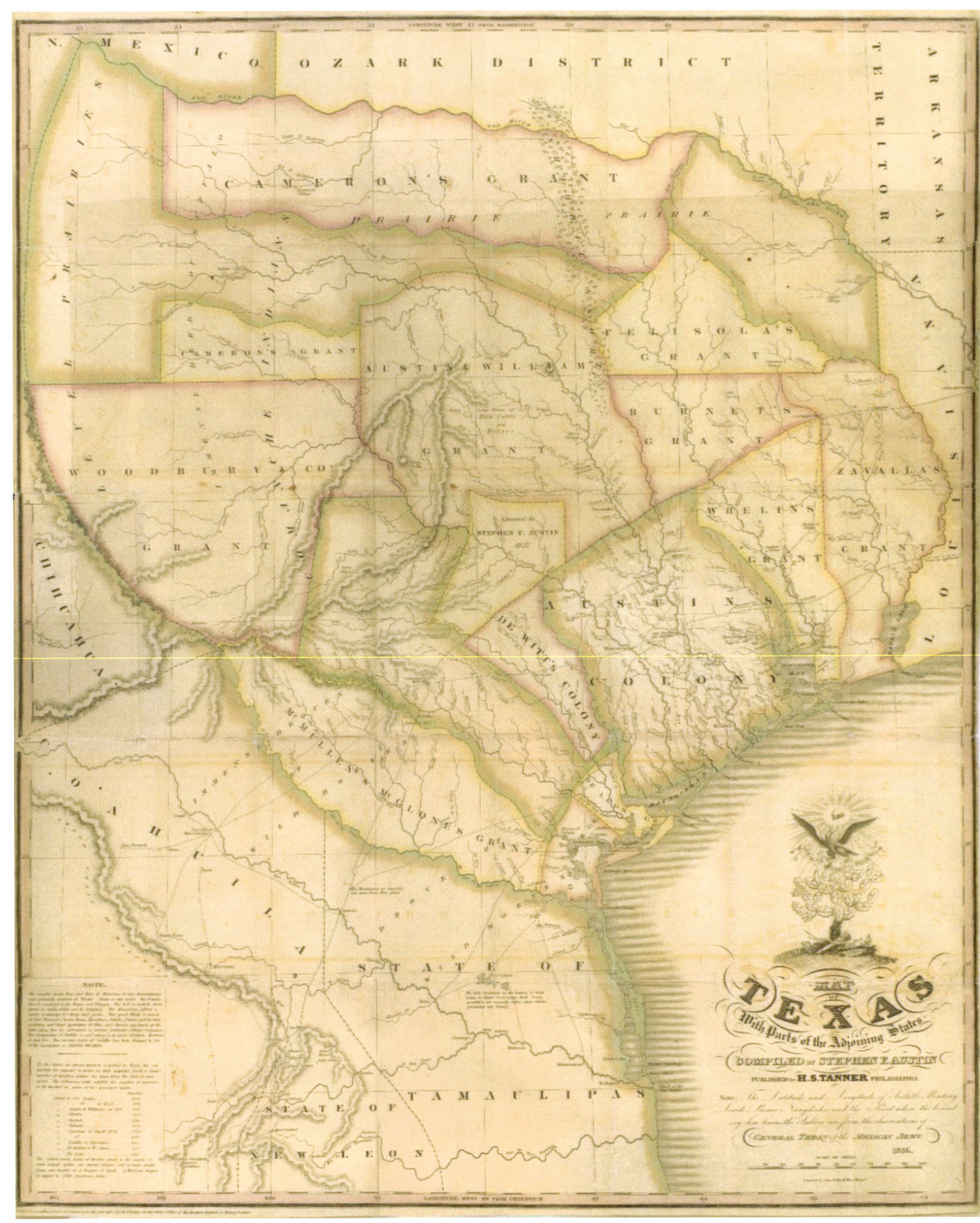

1836 version of "Map of Texas With Parts of the Adjoining States" compiled by Stephen F. Austin, published by H. S. Tanner of Philadelphia, with observations of General Terán of the Mexican Army.
An early published map of Texas, versions exist from 1830 to 1840. Austin cannily emphasized Texas despite being "Coahuila y Tejas" at the time, & featured symbolism above title for Mexican federalism.

Barranco Colorado
A Mexican Military Post along the Lower Lavaca River (1830-1832)

... an illustrated narrative by
Chris Kneupper

... published in cooperation with
Jackson County Historical Commission

Library of Congress Catalog Number 2025925358

International Standard Book Number 979-8-9932899-4-6

Copyright © 2025 by Chris Kneupper
All rights reserved
Printed in the United States of America
First Edition, Privately published
First Paperback Printing, 2025

DEDICATION

Dedicated to the memory of my father
Melrose Kneupper (1927-2021)
Navy veteran
"to go fight Kamikazes on a destroyer"
and the hardest working man I ever knew

ACKNOWLEDGEMENTS

The vast majority of information about Barranco Colorado was derived piecemeal from individual letters and dispatches in the Béxar Archives found at the Dolph Briscoe Center for American History (CAH), whose staff was extremely helpful in allowing the author to gradually assemble copies of microfilm and photo-duplicates of originals over the last several years. The Texas General Land Office was also most helpful in locating old documents in their Spanish Archives collection. Frank Condron and William Reaves of the Jackson County Historical Commission, Dr. Robert W. Shook of Victoria and Gary Ralston of Calhoun County provided local information and encouragement. Michael Bailey of the Brazoria County Historical Museum (BCHM) and the staff of the Brazoria County Library System were most helpful in locating books and other documents bearing on this topic. The librarians of the Brazoria County Library System (especially the Brazoria branch) helped me obtain numerous obscure titles through Interlibrary Loan. Unless otherwise stated, the author made transcriptions and English translations of many of the handwritten Spanish-language documents obtained from these sources, with the assistance of James E. "Jake" Ivey, Xavier Sendejo, Sonia Bennett, Flor Leon, Paul-Michael Dusek and Gregg Dimmick. Many of the images used in the Figures are photographs taken by the author or obtained from the Internet under Title 17 U.S.C. Section 107. Last but not least, my family (wife Helen, daughters Elissa & Christiane, son Carl) have provided endless support through years of archaeological projects and research, and here with digital editing of images, formatting, pagination, and proof-reading. The author expresses gratitude for all of this invaluable help.

However, this story would not be possible without the previously under-used resource of the Béxar Archives, due the fact that its documents of this era have not yet been translated, and their Calendar is a useful finding aid, although the Calendar is no longer maintained on-line by the Briscoe Center (instead use Wayback Machine link below).
https://web.archive.org/web/20160820043615/http:/www.cah.utexas.edu/projects/bexar/calendar.php

In general, we also owe a great debt to all the other here-unnamed historians, archivists, cartographers, authors, artists and history buffs who have created collections (such as the Béxar Archives, Austin Papers, Portal to Texas History, etc.), publications, drawings or maps, making it possible for us to build this illustrated story.

Dios y Libertad

Don Christiano

AUTHOR'S PREFACE

Have you ever heard of Barranco Colorado? Well, I hadn't, until a few years ago! You see, being a lad of the Brazos bottoms, I became involved in researching the 1832 Fort Velasco at the mouth of the Brazos, and found that old Velasco had a whole lot of history beyond that. So, I just had to tell that under-told story, eventually writing my first book "Here Rests the Brave – A Chronological and Archaeological History of the Forts Velasco". In that effort, untranslated documents in the Béxar Archives (maybe 8 in number) were key in understanding some of the details and timeline in the construction of that first *Fortaleza de Velasco* of Mexican Texas. A few others of these letters and dispatches came to or from some guy named Arteaga at a mysterious place called Barranco Colorado. Priding myself as a bit of an archaeology and history buff, I was a tad embarrassed to say I knew nothing of these names. The mystery got to me and, as I was finishing up the Velasco research, I begin to look into Barranco Colorado. I did the usual "lazy researcher" thing – examined Wikipedia and the on-line Handbook of Texas – but nothing! I then found a few obscure articles (secondary references) about it, most importantly a passage in a book by John J. Linn. I learned that Barranco Colorado had actually been a sister fort of the 1832 Fort Velasco, actually one of six that Manuel de Mier y Terán had tried to build in 1830-1832. And then, Holy moly! Those same Béxar Archives had over 100 documents to or from Aniceto Arteaga or his command, and I'd just opened a can of worms! Many were actual monthly military reports, quite unlike anything at Fort Velasco. I really didn't need a whole new job during my peaceful retirement but, what the hell, what else did I have to do during the COVID madness! So, I slogged through translating many of these documents. The roster lists didn't need much translating anyway! So, here is another under-told story from the days of Mexican Texas. Turns out, these efforts by Terán (an intelligent and insightful man) were his desperate attempt to assimilate Texas into the Mexican nation, but they failed in large part due to lack of proper resources from the new struggling nation of Mexico, often beset by revolution. Indeed, the failure of these half-hearted measures probably quickened the Texas Revolution of 1835-1836, the opposite of what Terán intended. Poor man, these circumstances caused his suicide in 1832. What an alternative he would have made, to Texas and Mexico, instead of the brutal Antonio López de Santa Anna!

> *Not to know what happened before you were born is to be a child forever. For what is the time of a man, except it be interwoven with that memory of ancient things ....*
>   - *Marcus Tullius Cicero, 46 BC*

TABLE OF CONTENTS

DEDICATION iv

ACKNOWLEDGEMENTS v

AUTHOR'S PREFACE vi

TABLE OF CONTENTS vii

LIST OF ILLUSTRATIONS viii

PURPOSE 1

INTRODUCTION 1

BACKGROUND 4

SECONDARY REFERENCES 19

DETAILS FROM PRIMARY REFERENCES 23

CONCLUSIONS 49

RECOMMENDATIONS 50

REFERENCES 50

INDEX 61

LIST OF ILLUSTRATIONS

Figure	Current caption used in narrative	Page#
	Cover art	i
	1836 version of "Map of Texas With Parts of the Adjoining States" by Stephen F. Austin	ii
	Melrose Kneupper	iv
1	Portion of original (1830) version of Austin and Terán ma	1
2	AI-generated image of Manuel de Mier y Terán	2
3	Mexican Forts of 1830-1832 (based on the 1836 version of the Austin and Terán map)	3
4	José de Evia's survey map of Bahía (& Lago) de San Bernardo	6
5a	1799 map of Gulf of Mexico by Juan de Langara	7
5b	Detail from 1799 map of Gulf of Mexico by Juan de Langara	7
6	Detail from 1807 Spanish Admiralty map for the upper Texas coastline	8
7	Stephen F. Austin's copy of the Puelles 1807 map	9
8	Hand-drawn map by Stephen F. Austin, circa 1822	10
9	"A Map of the Country between the Brassos & La Baca Rivers", N. Rightor, 1822	11
10	1829 Berlandier map of "Bahía de San Bernardo ou Bahia de Matagorda"	13
11	Sketch of Aranzas Bay surveyed by Eugenio Navarro, 1832	14
12	1857 USCS chart of Entrance to Matagorda Bay	15
13	Inset of upper Lavaca Bay from USCS Chart No. 107, Matagorda Bay, Texas, 1872	16
14	Detail from USCS Chart No. 107, Matagorda Bay, Texas, 1872	17
15a	Dept. of Commerce & Labor, Bureau of Fisheries Chart of a Part of Matagorda Bay	18
15b	Detail from 1904-1905 Bureau of Fisheries Chart of a Part of Matagorda Bay	18
16	Portion of 1840 Map of Jackson County, showing Cox's Point	20
17	Advertisement about Cox's Point in The Texas Republican newspaper	21
18	Detail from "Victoria County" map, 21-Nov-1858 by Charles W. Pressler	22
19	Portion of Linn 1838 Connected Map of Victoria County	23
20	Notice in Texas Gazette, 22-Jul-1830, Page 2, Column 1	28
21	Monthly "summary" report by Arteaga and Castillo from Guadalupe dated 1-Aug-1830	29
22	Monthly "summary" report by Aniceto Arteaga from Guadalupe Victoria, 1-Oct-1830	30
23	Monthly "muster" report by Castillo from Guadalupe dated 3-Aug-1830	31
24	Certificate of attendance for Aniceto Arteaga at commissary review, by Rafael Chowell	32
25	Monthly "summary" report by Barberena from Goliad dated 3-Aug-1830	33
26	Commissary Review Report for Convicts by Trujillo at Guadalupe, 3-Sep-1830	34
27	Budget Report for 3rd Active Company by José Adeodato Vivero dated 1-Feb-1831	35
28	Military Population of Lavaca Detachment (with Grand Total)	37
29	Military Population of Lavaca Detachment (separated by unit identity)	37
30	Goods confiscated from the schooner **Hetta**	40
31	"Ruta de Goliad a Guadalupe (Victoria) y la Vaca", ascribed to Jean Louis Berlandier	45
32	Detail from 1839 Hunt-Randel map of "Texas"	46
33	Information from TARL database for 41JK29	47
34	Likely Location of Barranco Colorado using 1952 USGS Map as Basis	48
35	Portion of Plate 4 from 1938 USACE survey of Lavaca and Navidad River watershed	48

PURPOSE

It is the purpose of this document to organize and publicize information about the Mexican fort or post named as "Barranco Colorado", founded on the lower Lavaca River beginning in the summer of 1830, and labored over for a period of two years or so, until it was abandoned along with several other military posts in the eastern and southeastern part of Mexican Texas in the middle of 1832.

INTRODUCTION

As the 1830's dawned in southeast Texas, significant but mostly rural settlement had been underway for almost a decade in this previously undeveloped area, largely through the colony established by Stephen Fuller Austin known as Austin's Colony, with his original settlers known as the Old Three Hundred. The only towns of note were San Felipe de Austin, Brazoria, Matagorda and Harrisburg, each only a few years old, as shown in the original 1830 version of a map created and commissioned by the empresario himself (Figure 1 below). This map shows that the southwestern boundary of Austin's Colony was the Lavaca River.

Figure 1: Portion of original (1830) version of Austin and Terán map published by H. S. Tanner

A Mexican general officer, Manuel de Mier y Terán, visited Texas as leader of a boundary-commission expedition and inspection tour (*Comisión de Límites*) from late 1827 to early 1829, visiting Laredo, San Antonio de Béxar, Gonzales, San Felipe de Austin, Nacogdoches and the east Texas border area (boundary line set by the Adams-Onis Treaty of 1819), before returning via the Coushatta Trace through San Felipe, Guadalupe Victoria and La Bahia to Matamoros [Morton 1944, Terán 2000]. Terán was considered *"... one of the most admirable men of the Mexican revolutionary era ... a brilliant tactician, a broadly interested scholar, a sympathetic leader, and an outstanding patriot"* [Berlandier 1980 p. xii]. After his visit to Texas, and alarmed at what he had seen, Terán became one of the advocates for a revised immigration policy and stronger military presence, later writing an influential report about his visit that was issued in early 1830. After playing a pivotal role in repelling a Spanish expeditionary force at Tampico in Aug-1829, Terán was promoted to "General of Division" with the post of Commander General of the Eastern Interior Provinces (which included Texas), eventually establishing his headquarters at Matamoros in Mar-1830.

Figure 2: AI-generated image of Manuel de Mier y Terán

In this role, Terán initially had plans to gather a large military force at Matamoros to be used in Texas as necessary [Morton 1944 pp. 194-196]. Stephen F. Austin, hearing of these plans, published a notice and editorial in the *Texas Gazette* in an attempt to assure his colonists this was in their best interests [Austin 13-Mar-1830]. However, these plans were altered somewhat by a new law soon enacted by the Mexican federal legislature.

Based on Terán's report, Lucas Alamán (Mexican minister of foreign relations) and other Mexican politicians created the infamous Law of 6-Apr-1830, in some cases exceeding Terán's advice. One provision called for the military occupation of Texas using, in part, convicts as laborers and soldiers. Another important aspect of the law was that authority for colonization in frontier states was vested in federal commissioners, removing such authority from the individual states. This last provision was in direct opposition to Stephen F. Austin's stated opinions [Austin 29-Mar-1830]. For Texas, the post of colonization commissioner was added to Terán's duties in late Apr-1830 [Morton 1944 p. 199]. The new law also forbade further immigration from the United States, while sanctioning further immigration from Mexico and Europe. Another provision of the law was Article 12, which stated *"Coastwise trade shall be free to all foreigners for the term of four years, with the object of turning colonial trade to the ports of Matamoros, Tampico and Veracruz."* [Howren 1913 p. 416]. This law, justified from the Mexican government's perspective, had a negative and galvanizing effect on the loyalty of the Anglo-American colonists in Texas (Texians), and its effect is often equated with the "*Stamp Act*" in catalyzing the American Revolution. However, this law's immediate effect was to give birth to efforts for new military sites in Texas, to enforce its customs and immigration provisions.

In his new position, Terán planned to construct a series of new forts in Texas, with one being near the mouth of the Lavaca River, apparently with the intention to control customs and immigration into the "Bahía de San Bernardo", later known as Matagorda, Lavaca, and perhaps Espiritu Santo Bays. This fort, and two others, were among the first actions of Terán in compliance with the new law, as indicated in a letter he later wrote to José María Viesca, then governor of the Mexican state of Coahuila y Tejas *"… I have selected some points which have appeared to me the most appropriate for locating detachments of troops: at the mouth of the Lavaca River, at the crossing of the Brazos River on the Upper Road from Béxar to Nacogdoches (a place which has been given the name of Tenoxtitlan), and at the point at the head of Galveston Bay, on the left bank at the mouth of the Trinity River."* Viesca responded with full approval, authorizing Ramón Músquiz (then, political chief in Texas) as commissioner to build the establishments [Terán 30-Dec-1830]. In total, six new forts were to be constructed to enforce the new law, including these three (Barranco Colorado, Fort Tenoxtitlán, Fort Anahuac), and also Fort Terán, Fort Lipantitlán, and Fort Velasco, adding to existing garrisons at San Antonio de Béxar, Presidio La Bahía, and Nacogdoches (as illustrated in Figure 3), along with a war frigate to serve the coastal forts [Filisola 1848 pp. 65-66]. Although Terán requested such a vessel, also to act as a coast guard, it was never obtained due to lack of funds [Morton 1944 pp. 499-500]. Some of these forts also included smaller satellite posts, often as a temporary camp but also something more substantial (for example, customs houses on Galveston Island and the mouth of the Brazos as adjuncts to Fort Anahuac).

Figure 3: Mexican Forts of 1830-1832 (based on the 1836 version of the Austin and Terán map)

Unlike the other new forts, the one on the Lavaca River is poorly known in modern references, with only a few obscure accounts mentioning it. In his memoirs written some years later, Vicente Filisola wrote *"The town of San Felipe de Austin is the capital and focal point for the colonists of Texas. General Terán has ordered the occupation of this and El Paso del Caballo, Lavaca or Barranco Colorado, Fort Velasco, Galveston, Anáhuac, Fort Terán, Harrisburg, Nacogdoches and Tenoxtitlán and some others"* [Filisola 1848 p. 139]. This account seems to indicate that the name of the place on the Lavaca was "Barranco Colorado". The term "Fort Lavaca" does not seem to have been used at the time, although this term was repeatedly used for Barranco Colorado in a recent historical/archaeological report for Fort Lipantitlán [Jackson et al 2006] and a book about the old road paths in the area of Victoria county [Shook 2007]. The term "Fort Lavaca" is otherwise found as a name for a Civil War era battery at the current location of the town of Port Lavaca.

BACKGROUND

Lest one think that Barranco Colorado was the first attempt to establish a detachment, fort or port on the middle Texas coast, our story needs to start a little earlier, so as to get some understanding of prior efforts at and near Matagorda and Lavaca Bays.

Spanish explorers had intermittently visited the middle Texas coast a number times by sea and land, most notably during the period of the late 1680's when LaSalle had established Fort St. Louis about 5 miles up Garcitas Creek from its mouth on Lavaca Bay. The presence of this colony caused great focus and attention to fall upon Matagorda Bay, resulting in no less than eleven Spanish expeditions by land or sea to find and destroy it [Weddle 1999]. These expeditions caused the drawing of several early maps of the area, where Matagorda Bay loomed large, perhaps overly so, in both Spanish and French efforts to understand the geography of the huge wilderness, now threatened by French incursion, that Spain had claimed but largely ignored for 150 years. Indeed, the second (or Rivas-Iriate) maritime search from Veracruz arrived at the bay on 2-Apr-1687, sometimes called Bahía del Espiritu Santo since Cabeza de Vaca's previous naming of it. Their chief pilot, Juan Enríquez Barroto, re-named the bay as Bahía de San Bernardo for the first time [Weddle 1999 p. 99, Shook 2007 p. 86]. Both names would be variously used for many years afterward. At the entrance of the bay, he also named the leeward (west) point as San Francisco and the windward (east) point as Culebras [Shook 2007 p. 86]. Two days later, they came across the wreck of **La Belle**, inside the bay three leagues east-northeast of the entrance against the windward shore [Weddle 1999 p. 99, Shook 2007 p. 87].

Some notable maps from the period (although not illustrated here) can be observed via the link or at the listed references:
1. Carlos de Sigüenza y Góngora map of 1689 - Camino que el año de 1689 hizo el governador Alonso de León desde Cuahuila hasta hallar cerca del Lago de S[an] Bernardo el lugar donde havian poblado los Franceses [Martin 1982b p.16, Jackson et al 1990 p.49, Weddle 1999 Plate 8]
https://pares.mcu.es/ParesBusquedas20/catalogo/show/20911
2. Llanos-Cárdenas map of 1690 - Planta cosmográphica del Lago de San Bernardo, 1690 [Weddle 1999 Plate 11]
https://pares.mcu.es/ParesBusquedas20/catalogo/show/20914?nm

3. Alonso de León map of 1690 - Mapa del Viaxe que el año 1690 hizo el Gobernador Alonso de León desde Cuahuila hasta la Carolina [Bryan & Hanak 1961, Plate 7, especially inset of Matagorda Bay]
   https://pares.mcu.es/ParesBusquedas20/catalogo/show/20913
4. Marqués de San Miguel de Aguayo map of 1722 - Carta de la Bahía del Espíritu Santo de la Provincia de las Nuevas Philipinas, que dejó observada [Weddle 1999 Plate 12, Shook 2007 p. 187] – famous as the first known instance for use of the term "Matagorda" in regards to Matagorda Bay of Texas. It shows a drawing of a tree on the shoreline (perhaps near modern Port Lavaca), so perhaps it was an attempt to draw a single stout-trunked and/or wide-canopied specimen (or motte) of Live Oak trees on the bank, labeling it as "Fat Bush".
   https://pares.mcu.es/ParesBusquedas20/catalogo/show/2619286?nm or
   https://pares.mcu.es/ParesBusquedas20/catalogo/show/20957?nm or
   https://pares.mcu.es/ParesBusquedas20/catalogo/show/20937?nm
5. Le Maire map of 1716 – Carte nouvelle de la Louisiane et país circonvoisins dressée sur les lieux pour etre presentée a S. M^{te}. T.C. par F. Le maire pretre parisien et mission^e. Apostolique MDCCXVI [Jackson et al 1990 Figure 8 pp. 52-53, Figure 9 pp. 54-55]
   https://gallica.bnf.fr/ark:/12148/btv1b7912138r
6. DeLisle-Lemaire map of 1718 – Carte de la Louisiane et du cours du Mississipi dressée sur un grand nombre mémoires entrau.^{tres} fur aux de Mr. le Maire Par Guill.^{aume} Delisle del academie R.^{lo} des Seten [Martin 1982b p. 40, Jackson et al 1990 Figure 14 pp. 62-63]
   https://www.loc.gov/item/98685731/ or
   https://gallica.bnf.fr/ark:/12148/btv1b84912758/f1.item.zoom or
   https://texashistory.unt.edu/ark:/67531/metapth231385/

Barroto's place names such as Lago de San Bernardo, Punta de San Francisco, Punta de Culebras, Rio de San Marcos and Isla Blanca are retained in the first of these maps, as well as an indication for the wreck of **La Belle**. An early name for the Lavaca River was Rio de San Marcos, given by Alonso de León [Weddle 1999 p. 194]. The fourth-listed map (1722) uses the term "Matagorda" for the first time, for a place on the shore of Lavaca Bay near what would become the port of Linnville over 100 years later, and also shows a deep-water channel along the southwest bank [Shook 2007 p. 187]. The fifth and sixth maps are French, copying information about Bahía de San Bernardo gleaned from Spanish maps of the LaSalle era, and the work of Saint-Denis and Le Maire [Jackson et al 1990].

The first successful and comprehensive attempt to map the coastal geography of the Texas coast, though, had to wait for many years. It came with the survey efforts of José de Evia in the 1780's for much of the coastline of the Gulf of Mexico. Detailed maps were produced of any significant bay and harbor discovered, and this included "Bahía de San Bernardo" [De Evia 1785], as shown in Figure 4 below, and "Bahia de Galveston" during separate voyages in 1785 and 1786 [Weddle 1992 pp. 113-115]. Please note the survey included soundings for water depth, and that a deep-water channel was found extending into the bay, perhaps to the point known today as Port O'Connor and Indianola. The tip of the barrier island just west of the mouth of the bay was labeled as "Punta de San Francisco", and the island eastward was "Ysla de San Luis". Modern Lavaca Bay is labeled "Lago de San Bernardo", while modern Matagorda Bay is "Bahía de San Bernardo". A river is shown emptying into the north shore of Lavaca Bay labeled as "Rio Colorado ó de Cañas", which may be the river known now as the Lavaca

River. Unfortunately, these surveys of individual bays were not assembled into a larger and/or public map of the Texas coastline at the time, and the Spanish authorities kept them secret. A "Portulano de la America Setentrional" (a collection) of the individual 121 harbor maps was assembled [Taliaferro 1988], presumably for private use by Spanish mariners. Interestingly, as José de Evia had apparently divided his work into several annual seasons, they "missed" the area between Galveston and Matagorda Bays when stopping and then resuming work. Another publication of the "Portulano" in 1818 showed a similar map: https://www.davidrumsey.com/luna/servlet/detail/RUMSEY~8~1~335211~90103095:Bahia-de-S--Bernardo-

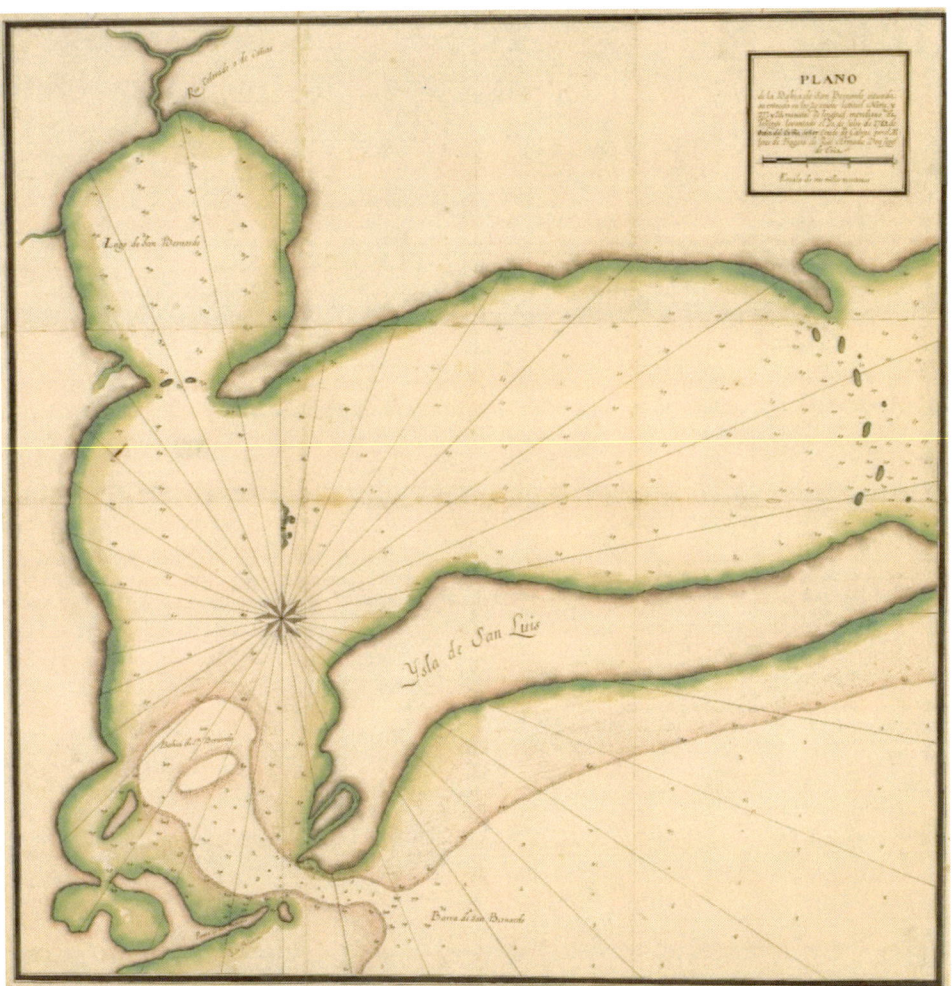

In 1799 and again in 1807, Spanish cartographers made large-scale maps of the Gulf Coast using the de Evia surveys, which retained this missing area, drawing it as if "Isla de San Luis" extended all the way from the mouth of Matagorda Bay to the entrance of Galveston Bay. Apparently, the Colorado and Brazos Rivers were missed entirely by de Evia, and an error (the drawing of Matagorda Bay connected to West Galveston Bay inside of a single barrier island) was made, which propagated into maps for decades afterwards. Curiously, this island (probably named for Fort St.

Figure 4: José de Evia's survey map of Bahia (& Lago) de San Bernardo

Louis) transferred its name to what later became known as Galveston Island and (on its western end) San Luis Pass. The map of 1799 was prepared by Juan de Langara of the Spanish Hydrographic Service, and had the ponderous title of "Carta Esferica que comprehende las costas del Seno Mexicano Construida De Orden Del Rey En El Deposito Hydrografico De Marina", which is shown below in Figure 5a in its entirety for an 1805 version [De Langara 1799]. A magnified portion of it for the upper Texas coast is shown in Figure 5b. Compared with de Evia's chart, the Lavaca River is labeled solely as "Rio Colorado" and the name of "Rio Flores" has been added to what might be modern Garcitas Creek.

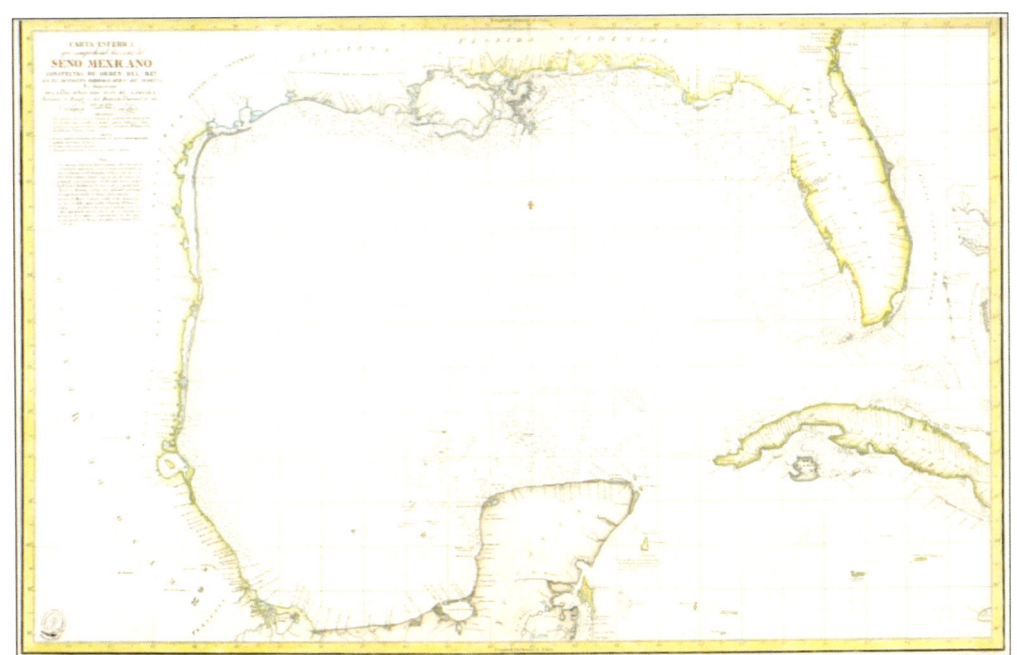

Figure 5a: 1799 map of Gulf of Mexico by Juan de Langara

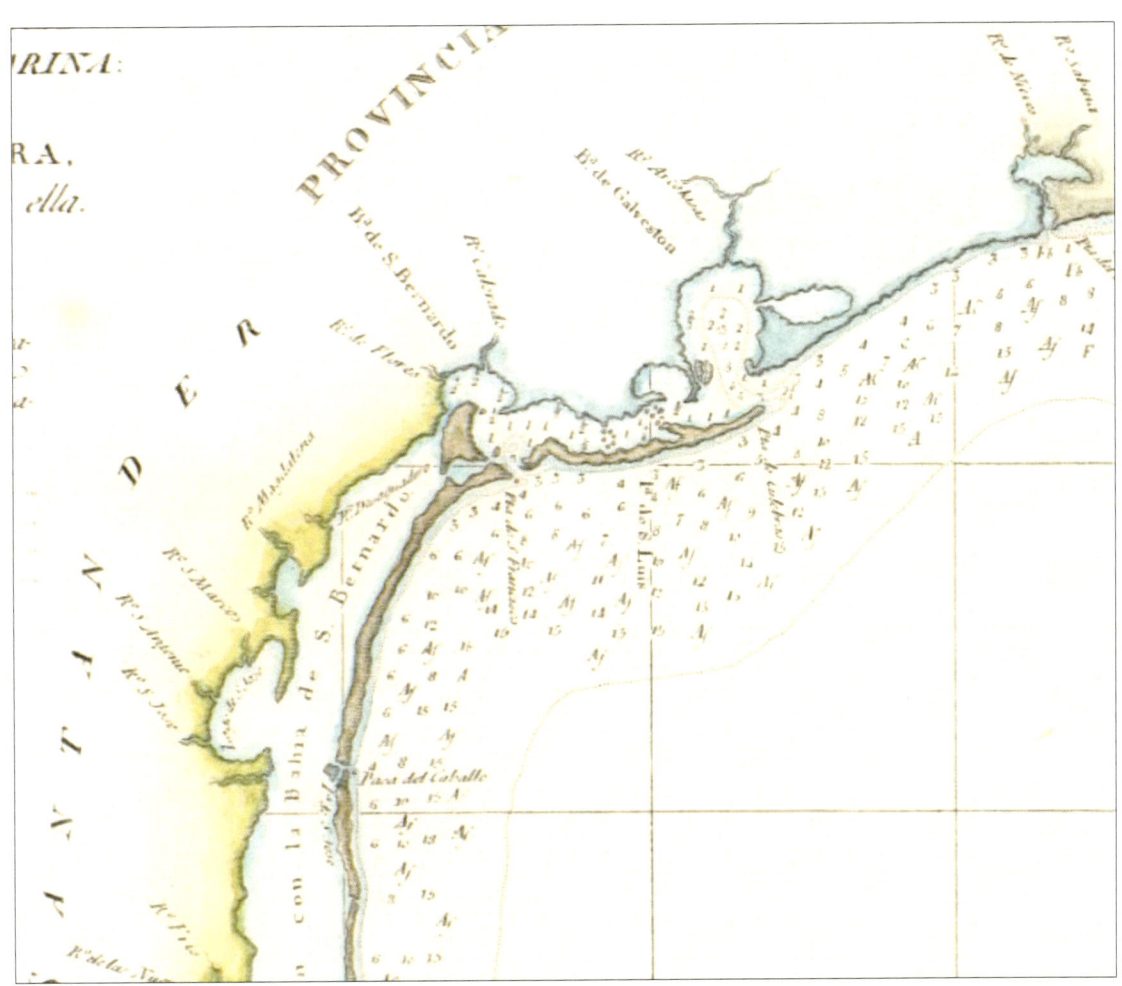

Figure 5b: Detail from 1799 Langara map for the upper Texas coastline

A similar map was prepared in 1807 by the Spanish Admiralty, which omitted the area of Florida, known as "Carta particular de las Cóstas Setentrionales del Seno Mexicano que comprehende las de la Florída Ocidental las Margenes de la Luisiana y toda la rivera que sigue por la Bahia de S. Bernardo y el Rio Bravo del Norte hasta la Laguna Madre" [Spanish Admiralty 1807]. A detail of the upper Texas coast from it is shown in Figure 6 below.

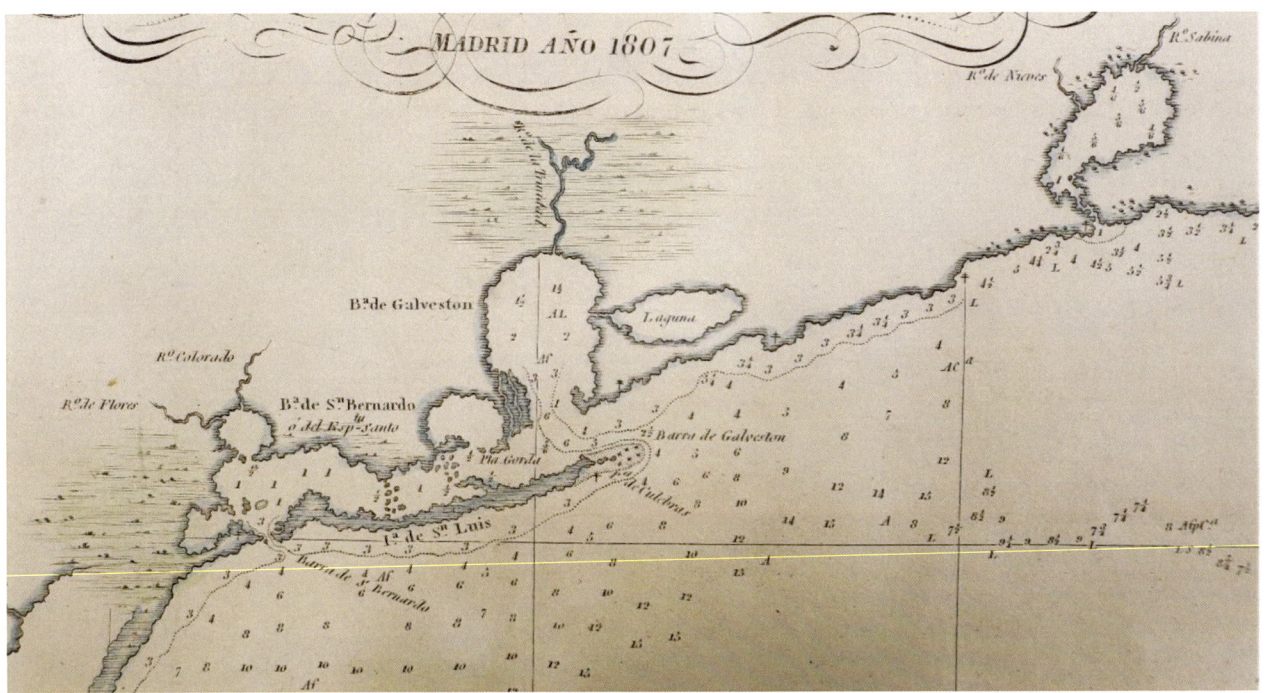

Figure 6: Detail from 1807 Spanish Admiralty map for the upper Texas coastline

Please note that this map also omits the area about the mouths of the Colorado and Brazos Rivers, and the single barrier island is again labeled as "Isla de San Luis". An area southwest of Matagorda Bay also seems to be intentionally left unclear.

Before Mexico won its independence from Spain in 1821, the King of Spain authorized Bahía de San Bernardo for free trade on 28-Sep-1805 [Soler 1805], and this was recognized again by the Eastern Interior Provinces of Mexico in 1821 [Barker 1926 p. 178-179]. This was probably due to the fact that these bays provided the closest deep-water harbor to the settlements, ranchos and missions along the lower San Antonio River near La Bahía (modern Goliad), and their incomplete knowledge of other harbor options along the middle and upper Texas coast. So, again, one can see that this area had been explored and known for some time. Nonetheless, *"No port worthy of the name was ever developed on the Texas coast during Spain's tenure."* [Weddle 1992 p. 118].

Some indication of the state of geographical knowledge of inland areas of the time can be seen in another 1807 map, drawn by a Franciscan friar at Nacogdoches, Fray José Maria Puelles Figure 7 is a digital copy of the Puelles map found among Stephen F. Austin's papers (the original is now at the Texas Map Collection at the Dolph Briscoe Center for American History – copy at Portal to Texas History - https://texashistory.unt.edu/ark:/67531/metapth298413/). This map was probably the best map of Texas for its time, although it, like many others, was never publicly released by Spanish authorities.

Published maps, such as the Humboldt map of New Spain (drawn 1804, published 1810 - https://www.loc.gov/resource/g4410.ct000554/) and the Zebulon Pike map of the Internal Provinces of New Spain (visited 1807, published 1810 - https://www.loc.gov/item/99446138/) were less accurate for the Texas interior and its rivers, although they were similar for coastal geography.

Figure 7: Stephen F. Austin's copy of the Puelles 1807 map; other versions are labeled "Mapa Geografica de las Provinciales Septentrionales de Esta Nueva Espana"

On these maps, the series of bays known today as Matagorda and Lavaca Bays is again labeled as "*Bahía (or Lago) de San Bernardo*", San Antonio/Aransas/Copano Bays as "*Bahía de San Jose*" and Galveston/Trinity Bays as "*Bahía de Galveston*". Galveston Island was labeled as "*Isla de San Luis*" and its northeast end was "*Punta de Culebras*", and the island was shown reaching all the way southwest to the entrance of Bahía de San Bernardo. With the exception of the Puelles map, the Brazos River is poorly represented as a minor river emptying into an intervening bay. With this state of knowledge, it is little wonder that Bahía de San Bernardo was considered the best choice for a Texas port. Like previous maps, a river labeled as the Colorado River is shown to empty into Lavaca Bay, although it in fact emptied into the northeast end of Matagorda Bay.

Perhaps, we can also infer why Moses Austin and his son Stephen F. Austin sought a grant of land to the east of the Spanish settlements and authorized bay, and southwest from Nacogdoches and the Spanish missions of east Texas, in what was a larger-than-mapped poorly-known and undeveloped "wilderness". In 1822, Stephen F. Austin prepared the first of a series of maps for the Austin Colony (a version found in the Library of Congress is shown in Figure 8), which continued to display poor knowledge of coastal geography, although it is rich with inland information about roads, Indian villages and names, and the extent of forested lands (in green). Many rivers are represented, flowing correctly to the southeast, but he seems to have left out naming the Brazos River [Martin 1982a, Reinhartz 2015]. The coastal areas are not much improved from the Puelles, Humboldt or Pike maps.

Figure 8: Hand-drawn map by Stephen F. Austin, circa 1822 (Library of Congress version)

An apparent copy of this map by or for a Mexican army officer, José Dominguez Manso, is found in the U. S. National Archives (https://catalog.archives.gov/id/12007750 - illustrated in [Reinhartz 2015] Figure 9) that was captured in the Mexican-American War, and another hand-drawn version by Austin can be found at the Briscoe Center for American History (CAH), both of which clearly label the Brazos River. Notably, these maps display the early unimproved roads of the time, and a detailed discussion of these roads can be found in a book by Robert W. Shook, a retired history professor at the University of Houston-Victoria [Shook 2007].

Still-another map was also prepared in 1822 by Nicholas Rightor for the area between the Brazos and Lavaca Rivers, held at the CAH (Figure 9 below). Again, no improvement of coastal geography is apparent, but there is accurate information about rivers and roads, as well as extensive "prairie" areas.

Figure 9: "A Map of the Country between the Brassos & La Baca Rivers", N. Rightor, 1822

The first (of four) land contracts to Stephen F. Austin extended from the Lavaca River on the southwest to the San Jacinto River on the northeast, bounded by the coast and the "El Camino Real" or "San Antonio Road" (between San Antonio de Béxar and Nacogdoches), the boundaries of which can be seen in the Frontispiece and Figure 1. The very first effort to actually bring colonists there involved the voyage of the schooner *Lively*, intended for "Bahía de San Bernardo"- at the time, the only authorized port in Texas. It sailed from New Orleans on or about 23-Nov-1821 with about twenty colonists and important supplies steering for the mouth of the Colorado River to meet Stephen F. Austin, but instead dropped them at the mouth of the Brazos River after a difficult month-long trip [Lewis 1899]. Upon returning to Texas on a second voyage with more colonists and supplies in 1822, the *Lively* was lost on Galveston Island, although the passengers were rescued and continued on to the mouth of the Colorado [Bugbee 1899]. Ships and colonists continued to arrive, and by the summer of 1824, most of the Old Three Hundred had arrived, and taken title to much of the prime property along the lower Brazos and Colorado Rivers. Stephen F. Austin foresaw the need for an authorized port, and wrote to the military commander of the Eastern Interior Provinces (which included Texas) on 27-May-1823, asking for authorization on several points, including a port of entry and authority to issue clearances for vessels [Austin 1823], apparently without success.

After the Mexican federal legislature passed a national colonization law on 18-Aug-1824 that forbade settlement in a 10-league band along the coast, Stephen F. Austin must have felt some urgency to legalize a port, as this "littoral reserve" of the new law reneged on that portion of his grant, and threatened his primary commercial and immigration connections by sea. He formally requested permission to establish *"el puerto de Galvezton"* in a petition also asking to extend his empresario contract to an additional 300 (then 500) families [Austin 1824, White 1839 p. 582]. Although the land contract was successfully authorized by the new state of "Coahuila y Tejas" on 27-April and 20-May-1825 [White 1839 pp. 610-613], the port was separately authorized in a modest decree by the federal legislature on 17-Oct-1825 [Arévalo 1829 p. 6]. This decree anticipated creation of a customs house (*aduana marítima*), but did not specify the location of the port. Thus, *"el puerto de Galvezton"* became the second authorized port on the Texas coast. At the time, this term seems to have applied to a broad area of the coast; in 1830, its new administrator, George Fisher, defined it to include *"... an extensive coast, from the Sabine River to Matagorda Bay ..."* [Fisher 5-Jun-1830].

After inspection and survey of Galveston Bay and Island, probably over 16 days in Feb-1826 using the rented sloop *Mexicana* and a rowboat, Austin realized the island was without timber or freshwater, subject to inundation, and isolated from the mainland [Austin 1826a, Martin 1982a p. 384] - so he favored the existing port at the mouth of the Brazos River [Austin 1826a, Austin Dec-1829, Barker 1926 p. 180]. Austin's survey resulted in an improved chart of Galveston Bay and Island [Austin 1826a], although the chart was forwarded on to the governor of Coahuila y Tejas in Saltillo, and from there to Mexico City - no surviving copy is known to exist. Since Austin had been asked earlier by Mexican authorities to seek boats for use by their detachments on the middle Texas coast, he then chose to purchase the two boats he had used in Galveston Bay, initially suggesting they be delivered to Balandra Point (along current San Antonio Bay) complete with sails and tools [Austin 1826b, Ahumada 1-Feb-1826]. Later dispatches reveal they were delivered to Sabino/Balandra Point in late July, and were to be used from a newly-staffed satellite post of La Bahia called "Matagorda" [Manchola 29-Jul-1826, frames 480-481] – probably on the western shore of Matagorda Bay near the location later known as Indianola or Port O'Connor.

Accompanying Terán's expedition of 1828 had been the naturalist Jean Louis Berlandier, who detoured from San Antonio de Béxar, starting to La Bahía on 25-Feb-1829. There he met the captain of the **Paumone** (probably **Pomona**); they traveled overland to Cópano (northwest end of Copano Bay) from which they departed by sea on 11-Mar-1829. They sailed southeast through the mouth of Copano Bay, then turning northeast they traversed Aransas Bay, Carlos Bay and Espiritu Santo Bay and out the entrance of Bahía de San Bernardo for New Orleans, returning on 13-May-1829 the same way [Berlandier 1980 pp. 390-408]. Apparently, during this trip, Berlandier acquired knowledge of the coastal geography and drew at least two maps, one of which is shown below in Figure 10, still indicating very poor conception of the local bays [Berlandier 1829]. Although some great detail about the entrance to Bahía de San Bernardo appears correct, the adjacent bays and rivers are badly inaccurate, especially the Brazos shown in the lower right corner. Lavaca Bay and the Colorado River are poorly represented in the upper right corner, as a small eastern extension of Bahía de San Bernardo, when it really attached to the northwest corner. Notably, the "Punta de San Francisco" of earlier maps is now labeled as "Punta de Matagorda", and the opposite point is labeled as "Punta de la Culebra". Obviously, however, local sailors knew well how to use these adjacent bays, and they must have been considered part of Bahía de San Bernardo.

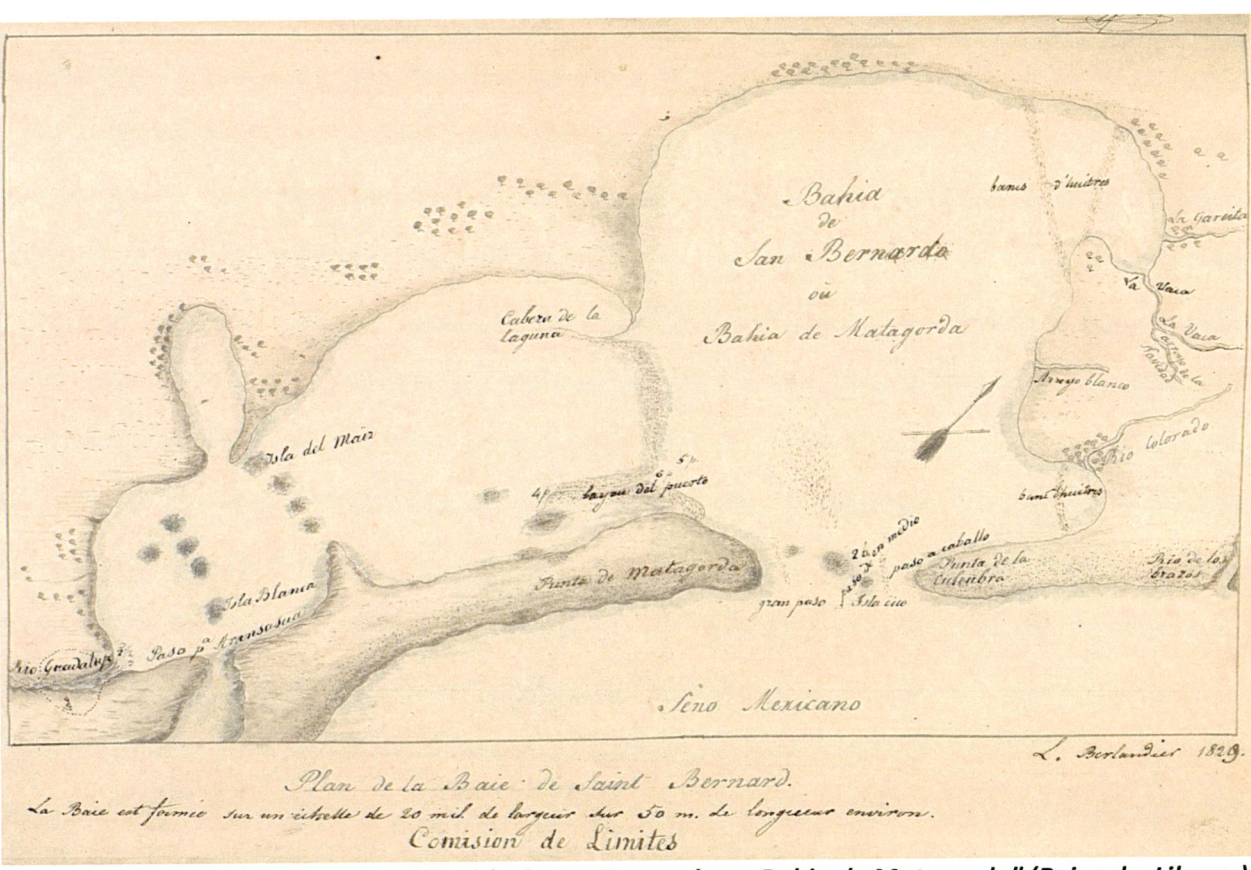

Figure 10: 1829 Berlandier map of "Bahía de San Bernardo ou Bahia de Matagorda" (Beinecke Library)

On 13-May-1829, the Congress of Coahuila y Tejas had issued Decree No. 94, setting conditions for customs officers to be placed at the ports of Galveston and Bahía de San Bernardo, including mention that the salary was $500 annually, and duties due to the state were to be two reales per ton [White 1839 p. 548]. José Antonio Navarro received an appointment from the government to become Administrator for the port of Galveston, but Stephen F. Austin advised him *"... that for some time it will not produce sufficient to live upon.*" [Austin Dec-1829], and he apparently never took such a position. So, as the 1830's began, it can be seen that geographic knowledge of the upper Texas coast was still very limited, but improving, and only with Stephen F. Austin's 1830 map (Figure 1) did an accurate widely-available representation occur. Apparently, Austin drew on information collected by the *Comisión de Límites* provided to him by Terán (such as Figure 10), to improve on his 1822 maps. For example, another Berlandier sketch of Matagorda, Lavaca, Espiritu Santo and Aransazu Bays, although faint (https://collections.library.yale.edu/catalog/2028634) bears great resemblance to those same bays drawn on the 1830 Austin/Terán/Tanner map (although still not completely accurate). Only with the 1839 Hunt-Randel map (https://libguides.uta.edu/ld.php?content_id=49096878) did a reasonably accurate representation of these lower bays occur. Tanner published several updates of the 1830 map through 1840, usually adding new towns and landmarks to the base map.

José Antonio Navarro's brother, José Eugenio Navarro, while an Alférez (2nd Lieutenant) in the Second Flying Company of San Carlos de Alamo de Parras, produced a sketch in 1832 of "Aranzas Bay" (today's Aransas and Copano Bays) and the navigable path to the early port of Cópano, today found in the records of the Texas General Land Office as Map #145 (see Figure 11 below). This may illustrate the path of the ***Pomona*** in taking Jean Louis Berlandier to and from New Orleans a few years earlier, and was one part of the improving knowledge of coastal geography.

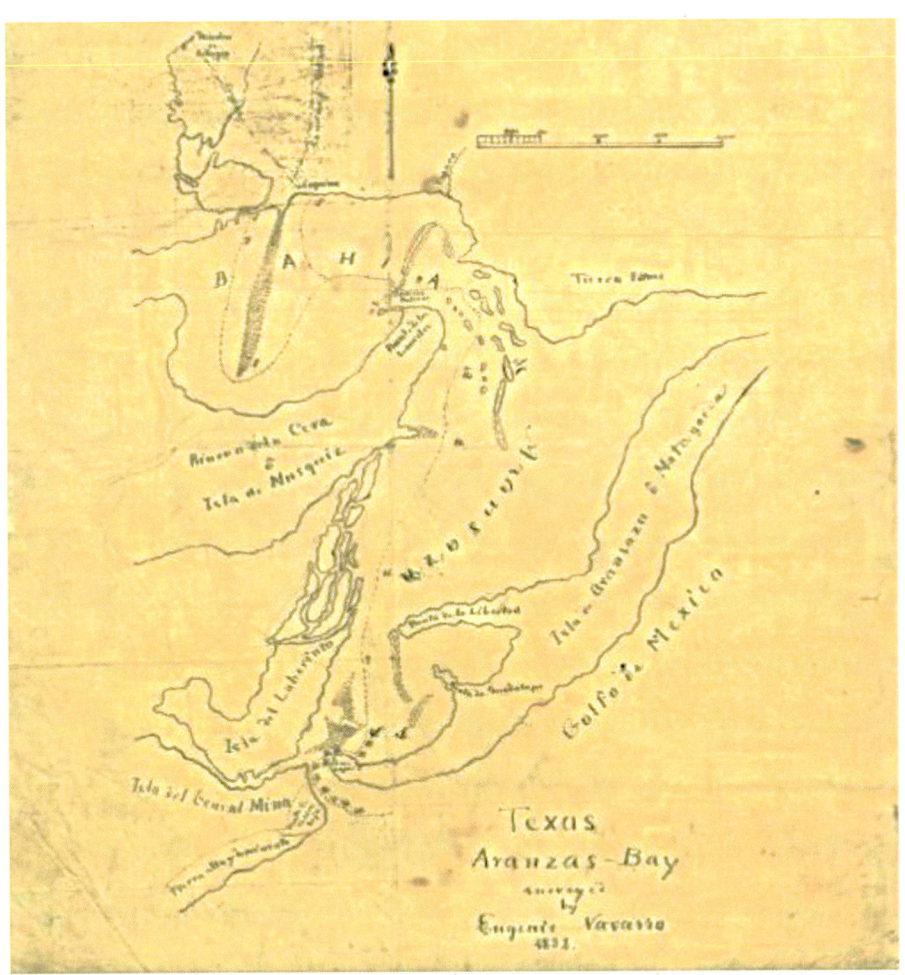

Figure 11 – Sketch of Aranzas Bay surveyed by Eugenio Navarro, 1832 (TGLO Map# 145)

Another version of this map (perhaps a later copy) can be found at Yale University's Beinecke Library, and is illustrated on page 239 of *Almonte's Texas*, a book published by Jack Jackson and John Wheat in 2003 [Almonte 2003].

If one examines Figures 4 and 10, it can be seen that a deep-water but narrow channel existed through Paso Cavallo into Matagorda Bay. Later, an 1857 chart by the U.S. Coastal Survey shows a more accurate rendering (left side of Figure 12). The point of land still attached to the mainland that is closest to Paso Cavallo is labeled as Alligator Head (later to become Port O'Connor), and the island just off that point is labeled as Bayucos Island.

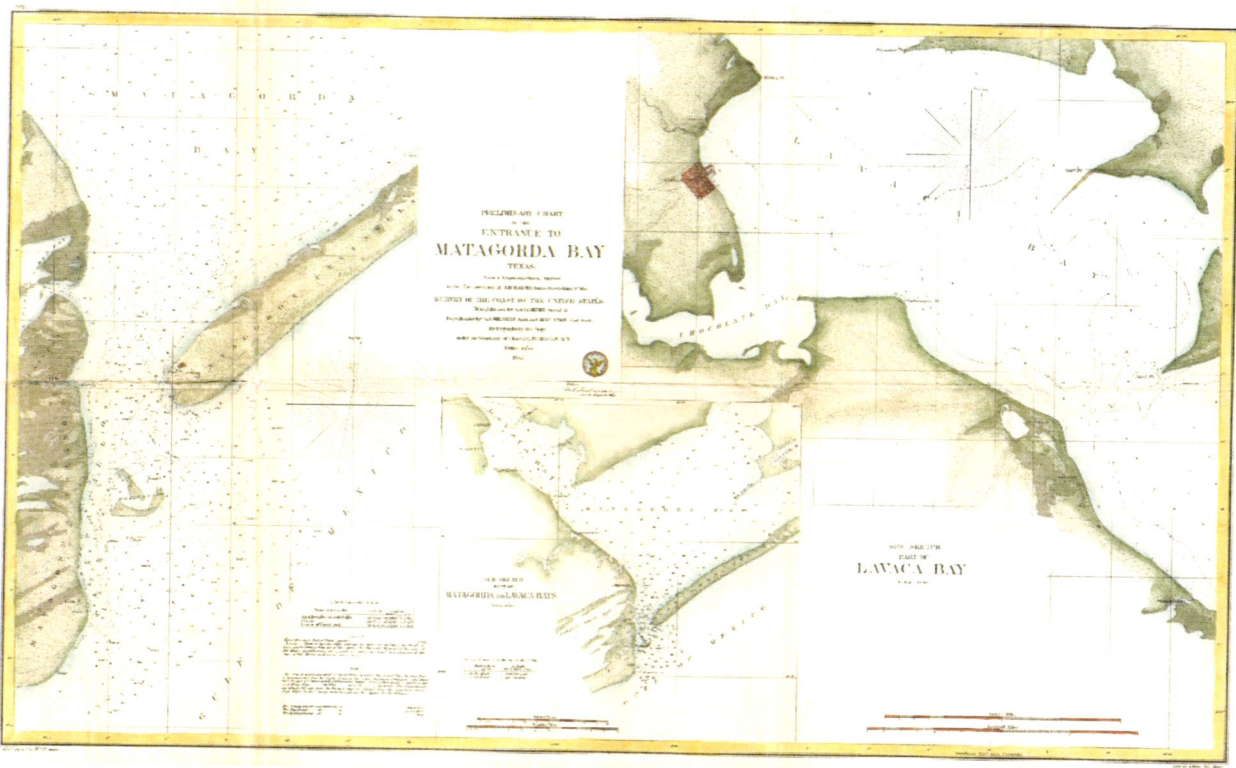

Figure 12: 1857 USCS chart of Entrance to Matagorda Bay [Bache 1857]

This map was updated in 1860 and again in 1872. The 1872 version is particularly instructive for the current study since it includes (for the first time) an inset for the upper lobe of Lavaca Bay. The older charts of Matagorda Bay seem to have neglected this portion of upper Lavaca Bay especially in regards to soundings; however, we find the 1872 USCS Chart No. 107 does have that type of information (Figure 13 below). One can see that 4 to 6' of water is reported in the northern head of the bay, reasonably close to Garcitas Cove and the mouth of the Lavaca River. Garcitas Cove itself appears very shallow, and there appears to be some spotty shallow reefs (oyster reefs?) in places in the bay. It is unclear, but there seems to be some kind of settlement between Garcitas Cove and the mouth of Venado Creek, at the shore near Bennett Point. Dimmit's Landing was reported to exist on the west bank of the Lavaca River at its mouth.

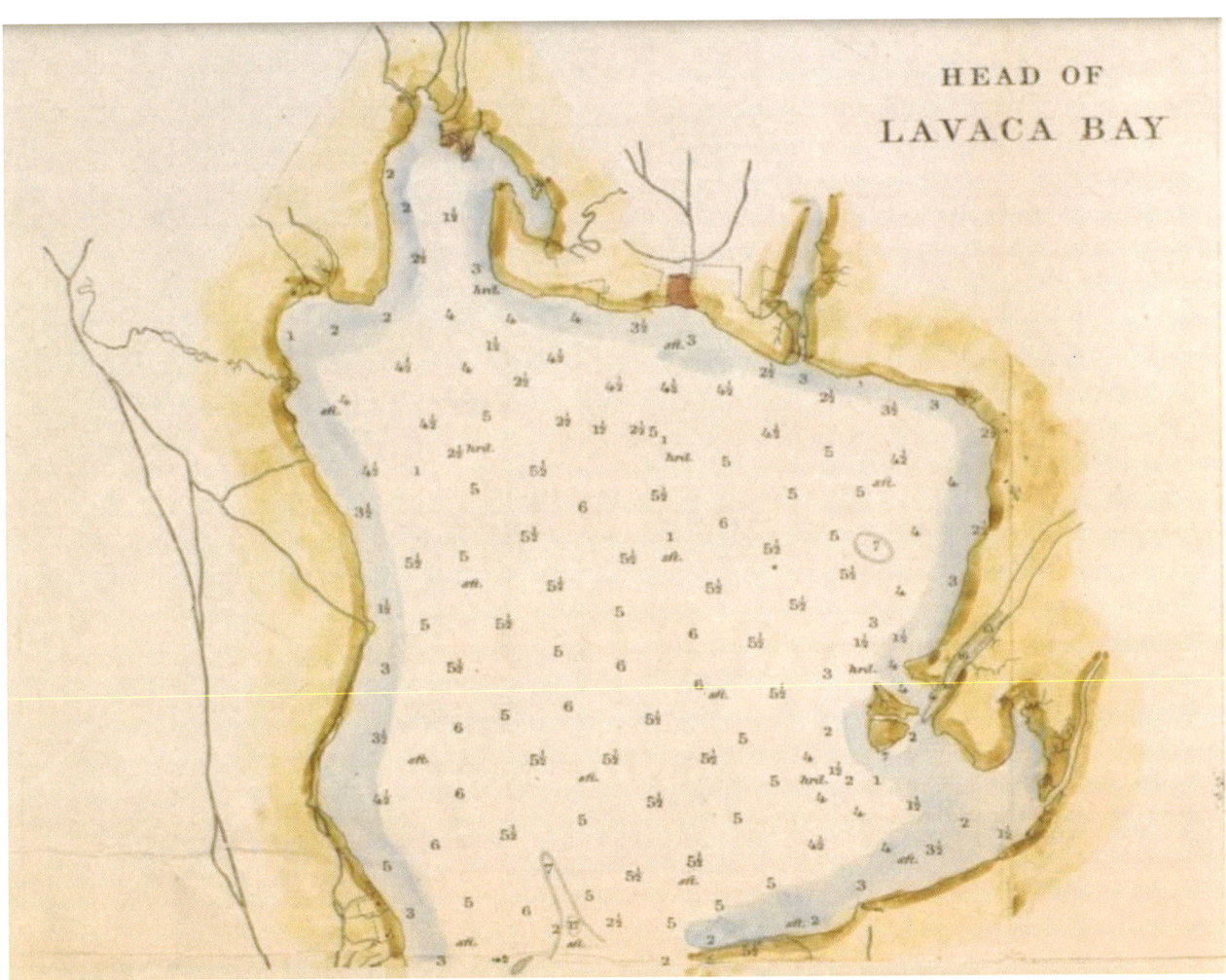

Figure 13: Inset of upper Lavaca Bay from USCS Chart No. 107, Matagorda Bay, Texas, 1872
courtesy of The Portal to Texas History (digital image 298357)

This same 1872 map also illustrates Matagorda Bay a bit further to the east, now including the town of Matagorda and the mouth of the Colorado River; see Figure 14. Notice that a reef (Dog Island Reef) and shallow waters block access to this area of the bay, although a finger of deeper water (of 6-7') does exist just inside Matagorda Peninsula between it, and Mad Island and Shell Island Reefs.

Figure 14: Detail from USCS Chart No. 107, Matagorda Bay, Texas, 1872
courtesy of The Portal to Texas History (digital image 298357)

Our review of the historic condition of *Bahía de San Bernardo* concludes with another map (Figure 15a), this one specifically drawn to chart the oyster reefs in the eastern portion of Matagorda Bay, in the period of Dec-1904 to May-1905. Ships of any significant size and draft would have to anchor off the end of Mad Island and Shell Island Reefs, and lighter any passengers or cargo for many miles.

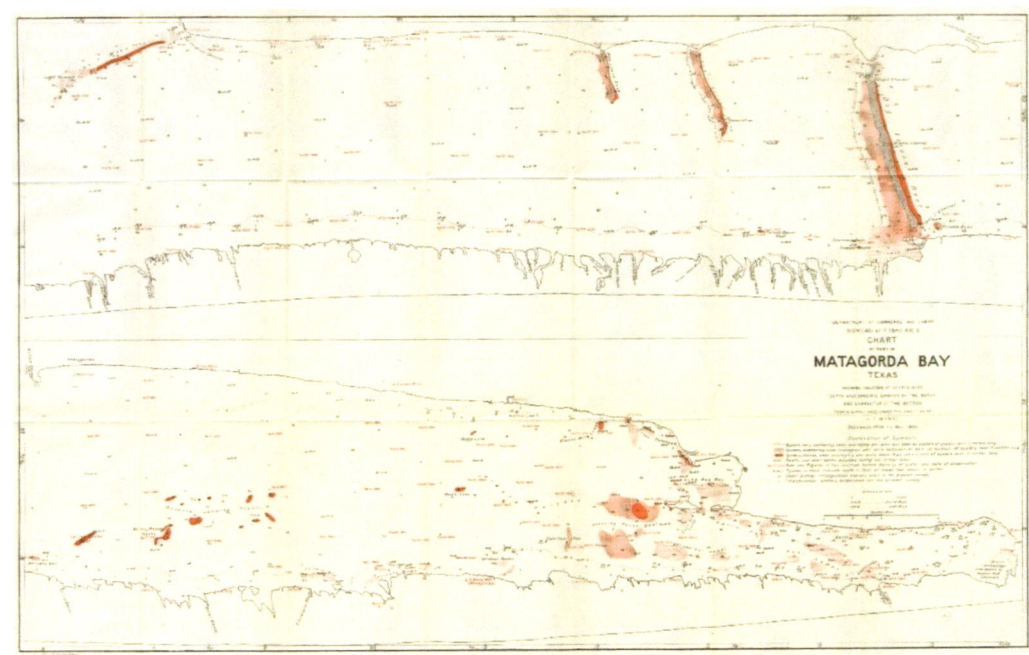

Figure 15a: Dept. of Commerce & Labor, Bureau of Fisheries Chart of a Part of Matagorda Bay
courtesy of Alamy Ltd.

Detail from the upper right portion (Figure 15b below) vividly illustrates the obstacles in reaching the mouth of the Colorado in historic times.

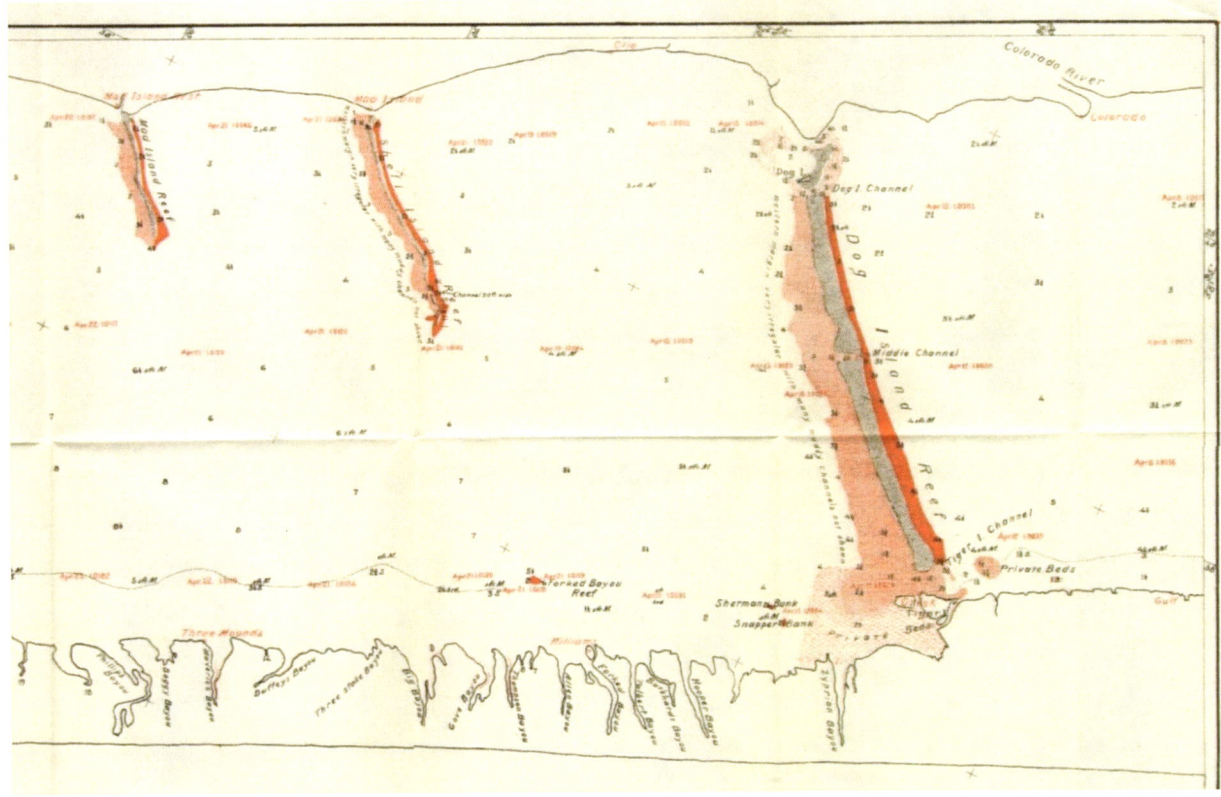

Figure 15b: Detail from 1904-1905 Bureau of Fisheries Chart of a Part of Matagorda Bay

As can be seen in the Figures 15, modern East Matagorda Bay was once part of a single larger Matagorda Bay. This situation remained until about 1929 when the Colorado "raft" was finally dislodged and a new delta formed into the bay. In 1934, a 200' wide x 9' channel was cut through this delta and across Matagorda Peninsula to empty directly into the Gulf. Thus, to reach the mouth of the Colorado in the early years, sailors had to enter *Bahía de San Bernardo* at Paso Cavallo, and then turn northeast to travel some distance up the bay to the mouth of the Colorado. However, at the time, the reefs in the bay, shallow waters and the "raft" prevented further navigation up the Colorado River itself much past (what became) the town of Matagorda [Clay 1949].

SECONDARY REFERENCES

An early settler of Jackson County, John S. Menefee (1813-1884), authored a series of articles in the **Jackson County Clarion** newspaper in the period of 20-May to 15-Jul-1880, a collection of which were transcribed many years later into a single typescript document entitled "Early Jackson County History". In this document is found a story alluding to a military post on the Lavaca (the context indicating it was the summer of 1831), stating "… *Capt. Mat (Nathaniel) Lewis and Capt. S. (Sylvanus) Hatch owned a vessel called the **Hetta**, by which father sent to New Orleans for some supplies, and he and I went down to get them on her return. Some Mexican soldiers from the Garrison on the west side of Lavaca went down also, and we camped at Cox's Point; next morning we and the officers went aboard of the vessel, leaving the soldiers ashore; the officers wanted the captain to send some water ashore for the soldiers, and after repeating their wishes two or three times the captain told the interpreter (Stoddard) to tell them to go to H__ell, which made the officers furious, they drew their swords and ___ the water went ashore and nobody was hurt, though somewhat scared. … The vessel was seized by the soldiers as having contraband on board, and lay in the bay until she became a wreck.*" [Menefee 1880]. The Béxar Archives has letters indicating this event occurred in the summer of 1831, although another interaction with the same vessel also occurred in 1830. The Spanish documents usually refer to this vessel as the **Hesta** (more information can be found later in this report).

The location of Cox's Point was a blunt peninsula near the northeastern corner of Lavaca Bay, just off the mouth of the Lavaca River, and can be seen on an 1840 map of Jackson County, in Figure 16 below (red arrow), showing a portion of this map surrounding Lavaca Bay. The map also shows the (unlabeled) mouth of Garcitas Creek at the northwestern corner of Lavaca Bay, the location of Linnville on the western shore (just north of modern Port Lavaca), the juncture of the Lavaca and Navidad Rivers and the site of the town of Texana (on the west bank of the Navidad River). Barranco Colorado would have been just west of Texana, on the west bank of the Lavaca River (red star). Modern Hwy-35 involves a causeway bridge across Lavaca Bay, and its eastern end is just north of where Cox's Point was located, now occupied by an industrial complex and harbor.

Figure 16: Portion of 1840 Map of Jackson County, showing Cox's Point (TGLO Map# 3708)

Cox's Point was later (in 1835) intended as a town site, as shown in an advertisement of the Texas Republican newspaper issue of 2-May-1835 of Brazoria, shown in Figure 17, and was the scene of some activity in the Texas Revolution. However, the 1840 map indicates no town at the site.

In his book "Reminiscences of Fifty Years in Texas", John J. Linn wrote *"... During this year (1830) General Teran sent two hundred soldiers to establish a military post on the Lavaca River. This addition to our scant population gave a decided impetus to trade. I engaged to supply the troops with all articles suitable to their wants."* [Linn 1986 p. 13]. At a later point in the book, he also wrote *"... There were in Victoria and on the Lavaca River above two hundred soldiers, who had been sent by the orders of General Teran in 1831. He had intended building a fort on the Lavaca; the definite position had not been decided upon, but the manufacture of brick had been commenced at a place called 'El Banco Colorado,' or 'The Red Bank' , on the west bank of the Lavaca and about three miles west of Texana. The works were under the control of Don Manuel Choval (Rafael Chowell?), a gentlemen of birth and education, commissioned by General Teran. ... The immediate commander of the troops was Captain Artiaga, also a*

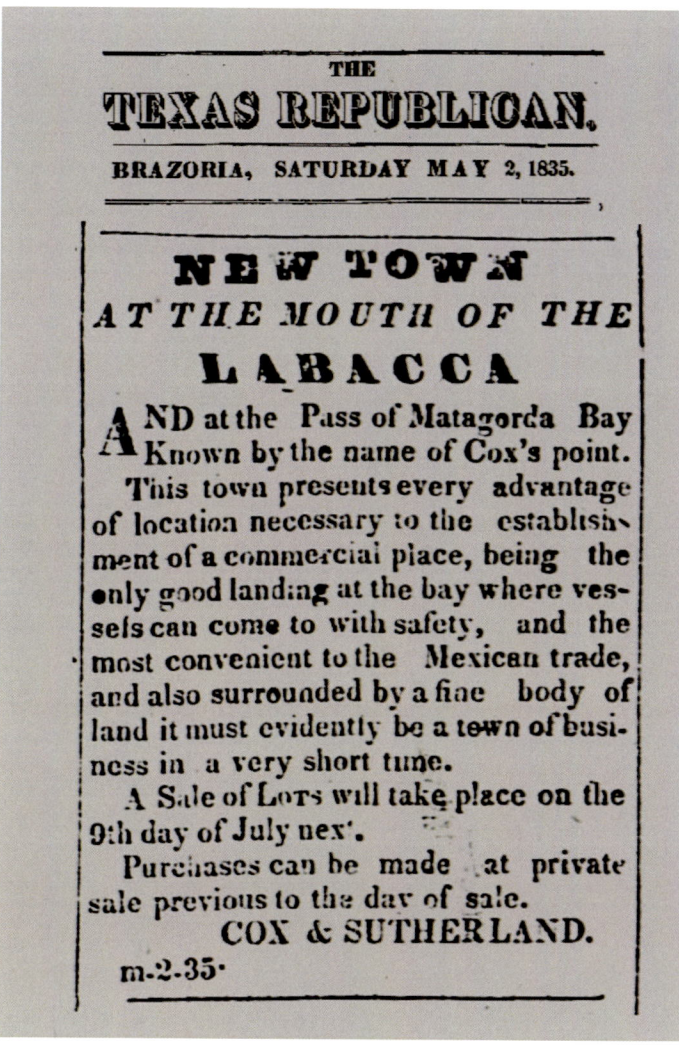

Figure 17 - Advertizement about Cox's Point in The Texas Republican newspaper

perfect gentleman and an old officer, who had served throughout the Mexican Revolution. ... After the surrender of Velasco *(late Jun-1832)* and the intrigues of Santa Anna had been developed, and after the death of General Teran *(3-Jul-1832)*, who committed suicide by falling upon his sword, Commissioner Choval resigned, and Captain Artiaga informed General Mexia that he wished to be relieved of the responsibilities of his position, as he did not favor the movements of Santa Anna. An order arrived directing the removal of the whole army, together with the workmen at the brick-kiln, some thirty or forty in number. These latter were from various parts of Mexico, who had been convicted of offenses against the laws and were known here as 'the chain-gang'. .... Lieutenant-Colonel Villasana arrived in a schooner in the bay to transport the troops to Matamoros, but had no funds with which to pay for provisions and other expenses. He had, however, authority to draw on the custom-house at Matamoros. ... As the 'Lavaca command' was indebted to me for supplies, Captain Artiaga called on me and stated that he was ordered to abandon the proposed fort; that he needed supplies, and that Villasana would draw on the Matamoros custom-house for the same. I knew their impecunious condition and the venture of accepting Villasana's check in payment, but so anxious were we all to get rid of the military that I determined to supply their necessities and run the risk of ever receiving pay, in order to contribute to the peace and prosperity of our country. All this was consummated. I supplied all their wants, and received of Villasana an order on the custom-house at Matamoros for some EIGHT HUNDRED and odd dollars. I sent the draft to a friend in Matamoros for collection. He was offered payment upon the basis of a ruinous discount which he did not deem at all equitable or just, and the proposed liquidation was rejected. .... But I have failed to receive one cent of that sum up to the present day. After the departure of the above-mentioned troops Texas enjoyed a period of peaceful quiet, absolutely free from the presence of Mexican soldiery, until the year 1835." [Linn 1883 pp.18-21].

Mindora Bagby McCallick (1906-1973), then a high school student, wrote an essay entitled "The Local History of Jackson County" to compete for the Caldwell Prize in Local History, which was judged by The University of Texas History Dept. Her essay, among others, was awarded a "Special Mention" and published in the University of Texas Bulletin of 22-Oct-1924, which said in part *"As late as 1832 the Mexican Government kept soldiers in a fort that stood on a high bluff on the west side of the Lavaca River. The traces of the fort and the old mission have almost disappeared and there are very few people living in the county at present who can locate the sites where they stood. At Dimmett's Landing on the Lavaca* (Dimitt's Landing was on the west bank near the mouth of the Lavaca River), *the Texans, in the revolution with Mexico, received many of their supplies and ammunition. About thirty-five years ago some of this land was put in cultivation and old cannon balls, bayonets, sabers, and muskets were plowed up."* [Bagby 1924].

Ira Thomas Taylor published the book "The Cavalcade of Jackson County" in 1938, with a small chapter entitled "Last Camp Site of Mexican Army in Jackson County". It says *"As late as 1831 and 1832 a Mexican army of some two hundred men with some thirty or forty convicts was stationed within the present boundaries of Jackson County, in the southeast corner of the John Linn Survey and on the west side of the Lavaca River near the present home of Charley Jones. It had been proposed by the Mexican Government to build a fort undoubtedly intended to overawe and if necessary to exterminate the citizens who had come from the United States in good faith to make their homes in Jackson County. … This military camp under the command of Captain Artiaga at a point called "El Banco Colorado" or the "Red Bank," was an army camp as well as a penal colony for convicts from old Mexico. The convicts were engaged in making brick and shipping them by boat to ports in old Mexico. Brick were actually manufactured there at that early date, and a number of such brick have been dug up from that brickyard and are now souvenirs of many of our citizens. … This army was removed and the camp abandoned by order of the Mexican Government in July, 1832. … All that remains of this old camp site on the high river bank is part of the remains of the old brick kiln."* [Taylor 1938 pp. 58-59]. The John J. Linn survey is shown in Figure 18 below, with a red arrow indicating the approximate location of Barranco Colorado, also showing an old road segment which appears headed toward it. The (northern fork) road segment labeled as "Texana Road" appears to be a portion of the trail from Guadalupe Victoria to San Felipe de Austin, although it is the southern fork that actually seems pointed towards Texana.

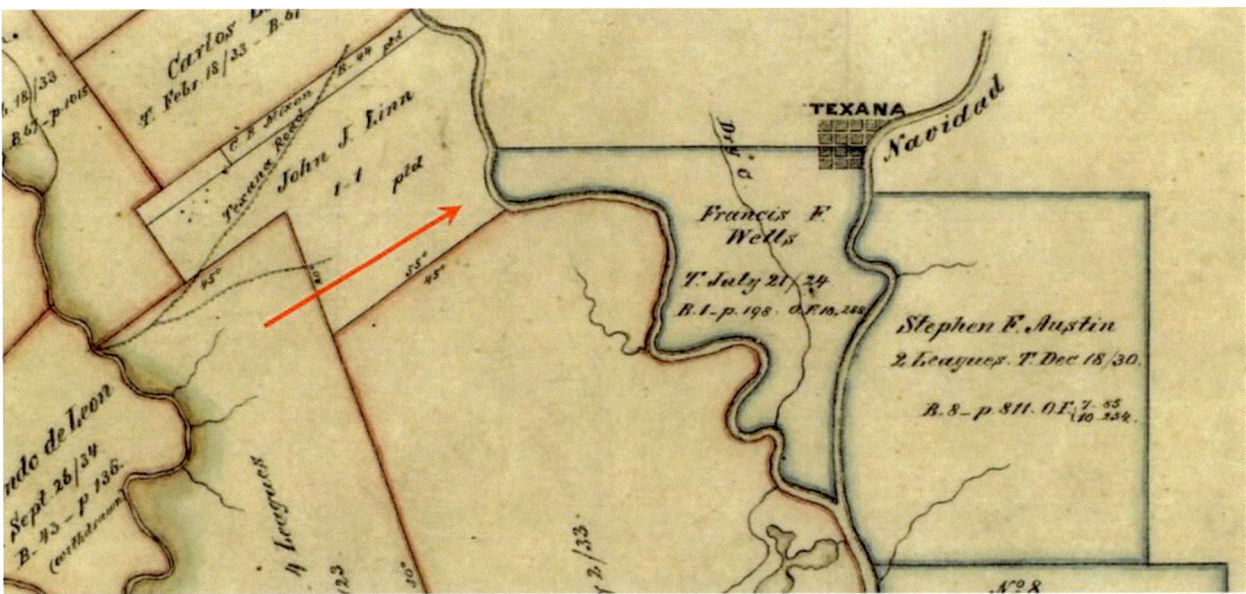

Figure 18: Detail from "Victoria County" map, 21-Nov-1858 by Charles W. Pressler (TGLO Map# 4115)

An earlier map created by Edward Linn in 1838 also shows these road segments, although it is in a degraded condition and is hard to read. In this map (a portion is illustrated in Figure 19), the southern fork is labeled as "Texana" (green oval) and the northern fork as "Hatch" (red oval). The Texana road crossing of the Lavaca River is shown to be just about the same location as the modern Jackson County Road 311 crossing, passing just south of the likely location of Barranco Colorado, although this map does not appear to show any indication of the post. In this period, the boundary between Jackson and Victoria counties was the Lavaca River, although the boundary later became Arenosa Creek, placing Barranco Colorado in modern Jackson County. This map also shows "Old Station Road" [Shook 2007 pp. 327-333], a reference to a temporary landing and settlement for the Green DeWitt Colony in the 1825-1827 period at the head of tidewater on the Lavaca River, known as Old Station (blue oval), which appears to have existed in the north corner of the league later granted to Leonardo Manso in 1834.

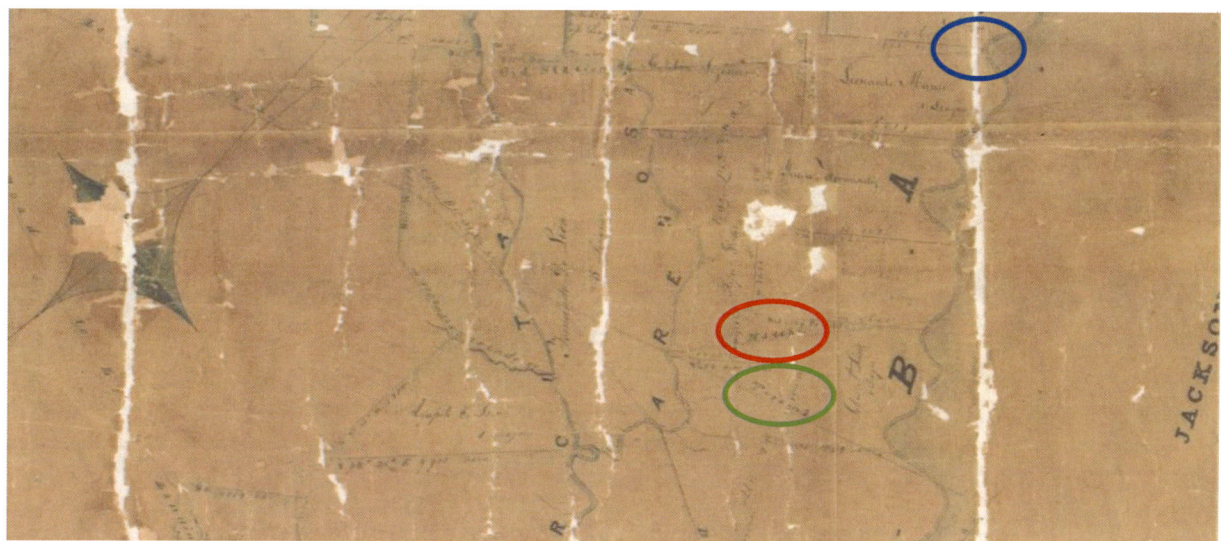

Figure 19: Portion of Linn 1838 Connected Map of Victoria County (TGLO Map# 1946)

DETAILS FROM PRIMARY DOCUMENTS

Now that some of the early mapping efforts have been presented, and a number of "secondary documents" reported that some kind of fort or military detachment existed on the lower Lavaca, can some "primary documents" verify the story, or provide more details of its construction, size or history?

A review of the Béxar Archives reveals over 100 letters to/from/about Aniceto Arteaga and the Lavaca post from 25-May-1830 to 16-Aug-1832 at locations shown as "Guadalupe" or "Barranco Colorado", perhaps indicating that (like Linn mentioned) some soldiers stayed in Guadalupe Victoria (modern Victoria, Texas) and some (perhaps convict laborers and their guards) on the Lavaca River making bricks for a fort, until early in 1831 when all moved to the fort. Many additional dispatches are found for coastal detachments and ship arrivals in the area just before and during the creation of Barranco Colorado. Review of the documents logged in the Béxar Archives Calendar for the latter half of 1832, though, indicate a likely error in recording the year of Arteaga's letters; almost all seem to properly be for 1831, since Arteaga wrote the numeral 1 much like a check mark, which has been confused with the numeral 2. Examination of his other letters indicates he wrote the numeral 2 in a different fashion, and this author believes no Arteaga letters exist beyond Aug-1832 about this location.

During this period, Arteaga's correspondence is labeled with a handwritten header and (later) a stamped letterhead as "*Comandancia Militar Del Establecimiento De La Vaca*", apparently a name applied to the command involving both Guadalupe Victoria and Barranco Colorado.

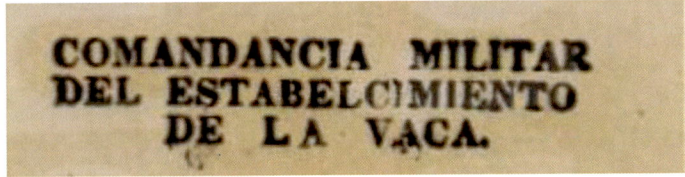 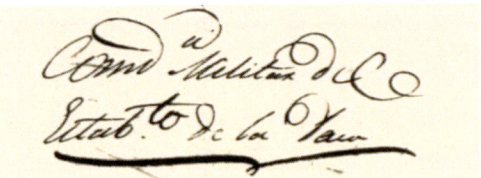

In early 1830, a number of dispatches reveal that a schooner named *Zorra* was being used by the port of Matagorda - apparently the *Mexicana* had been replaced in the period since 1826. A letter in February of that year reported that the *Zorra* had fallen into a state of disrepair and was "*useless*" [Cosío 26-Feb-1830, frame 710]. The boat was reported by its captain, George Midkiff, to have deteriorated after two years of use, but was being salvaged for parts [Elosúa 2-Mar-1830]. So, it appears that the port of Matagorda operated in some form or fashion up to about the time that Barranco Colorado was established.

Even before he had established his headquarters at Matamoros, Terán wrote from San Fernando to Antonio Elosúa authorizing the port of Matagorda to receive supplies for Texas troops, consigned to one Salvador Flores [Terán 25-Feb-1830]. Flores seems to have been acting as a civilian agent to receive supplies for the Mexican military posts. It is unclear why a civilian agent was needed, although Terán cited the Decree of 29-Sep-1823 which had authorized duty-free importation of any goods to Texas for a period of seven years. A sailboat or schooner called the *Oscar*, one such ship bringing supplies from New Orleans for troops in Texas in Feb-1830, went off course and was overdue, and then was apparently found and escorted by the schooner *Constante* into the Mexican port of Matamoros (Brazos Santiago), before being re-directed back to Matagorda in Apr-1830. However, it instead traveled all the way back to New Orleans, and the supplies for Texas had to be sent on another ship [Terán 24-Apr-1830].

During the Spring of 1830, a number of foreign ships called at the ports of the middle Texas coast, perhaps demonstrating by their example the need for an increased presence of Mexican authorities, such as that intended with Barranco Colorado. One of these was the merchant schooner *Sol*, Captain Pierre Boujean, who arrived from New Orleans at Cópano on 18-Apr-1830 with goods, provisions and passengers [Cosío 23-Apr-1830, frames 929-930]. It returned in a similar fashion again from New Orleans, arriving back at Cópano on 9-Jun-1830 [Cosío 2-Jul-1830, frames 149-150 & Elosúa 5-Jul-1830, frames 303 & 306]. Both trips presumably traveled via Aransas Pass as shown in Navarro's chart of Figure 11.

A schooner owned by Victoria merchant John J. Linn named the *Oposición* arrived from New Orleans at the anchorage at the mouth of the Lavaca River on 27-Apr-1830, with trade goods and John's two brothers, Edward and Charles Linn. The vessel was first reported by Sgt. José María de la Garza of the detachment at that point, observing the brothers did not have the proper passports, and also suspected smuggling of tobacco. Some drama ensued, when the Comisario of Goliad, José Bonifacio Galan, initially declined to travel to Lavaca to inspect the ship, since he insisted on a personal military escort. Eventually, Eugenio Navarro and José Valentin Garcia were then ordered to investigate, but found no evidence of smuggling [DeLaGarza 1830].

The *Oposición* returned to New Orleans and then Lavaca again on 27-Jul-1830, bringing three more passengers and 14 tons of cargo [Arteaga 25-Aug-1830]. The Linn brothers, unable to obtain valid passports, had to return aboard when this ship left Texas for New Orleans [Cosío 13-Aug-1830, frames 483-484].

A curious incident also occurred in this period involving the arrival of a "launch" from Matamoros. First reported when found abandoned on the Gulf beach south of Punta de Matagorda on 25-May-1830 by Corporal Francisco de la Garza, it was 19' long x 7' wide x 5' deep, had 5 oars per side, but needed caulking repair. Although events are not perfectly clear from the snippets of information in the many dispatches, it appears that a group of 10 armed Mexicans, 2 sailors and an American captain had come from Matamoros to the mouth of Matagorda Bay, claiming to be under orders of Terán and the captain of the port of Matamoros (Luis Asqueta), to commandeer all available merchant ships from Matagorda Bay to Corpus Christi for use back at Matamoros. Their "launch" was abandoned at Matagorda Bay, perhaps since it needed repair, and somehow, they acquired a small sailing vessel, taking the masts and sails from the damaged *Zorra*, and also Corporal José María Becerra (commander at the port of Matagorda), to travel on to Los Mosquitos (above modern San Antonio Bay). There, they encountered Mexican officers such as 2nd Lt. Eugenio Navarro who had been sent to investigate their claims, since they were suspected of really being corsairs or pirates. It is tempting to conclude that Terán and his subordinates did indeed send them, to gather vessels to support his efforts at colonization and fort-building since the government failed to fund the ships needed, although such direct evidence has yet to be found. The group then returned to Matamoros [Cosío 12-Jun-1830].

The abandoned launch was eagerly claimed by the port of Matagorda as a replacement for the *Zorra*. [Elosúa 21-Jun-1830, frames 823-824]. However, the captain of that port, George Midkiff, found that it would take substantial repair and cost, and Terán wanted the vessel for the harbor of Matamoros (Brazo de Santiago) and his wish apparently prevailed [Midkiff 1-Jul-1830].

The schooner *Hetta*, under Captain Nathaniel Lewis, arrived on 8-Jun-1830 from New Orleans with 57 settlers for Austin's Colony, first being spotted from the port of Matagorda by Francisco de la Garza. The vessel anchored in Lavaca Bay near the Garcitas anchorage, and was inspected by a small group including Fernando de Leon. After dropping the Austin colonists, the vessel apparently continued on to the Los Mosquitos landing, dropping a few other passengers and goods intended for Béxar. The vessel also seemed to have been targeted for seizure by Thomas M. Thompson, captain of a coast guard vessel based at Brazos Santiago, since it was suspected of smuggling tobacco. But, after Thompson's presence and actions ruffled the feathers of local authorities, Terán admonished this captain, ordering him to leave enforcement to the local authorities and to return to Matamoros [Cosío 17-Jun-1830]. A letter from Ramón Músquiz to Governor Viesca seems to provide the missing clue that it was Thompson, perhaps willfully misinterpreting his instructions to pursue tobacco smuggling, who arrived at Matagorda in the "launch", and then traveled on to Sabinito before being ordered back to Matamoros

[Músquiz 19-Jul-1830]. So, after abandoning the "launch", he and his crew may have boarded the **Hetta** on its passage back through the port of Matagorda on its way to Los Mosquitos/Sabinito.

The surviving records specific to Barranco Colorado and the new Lavaca post seem to start on 25-May-1830, when Terán sends a letter to Antonio Elosúa (also spelled Elozúa), the military commander for Coahuila and Texas based in San Antonio de Béxar, ordering him to assist creation of this post, and copying him on the instructions that Terán had given to the commander at Goliad (Mariano Cosío) and also to the unnamed commander of the war schooner **Constante** that same day [Terán 25-May-1830]. This was just one month after Terán had ordered the creation of Fort Tenoxtitlán, so the Lavaca post was the second of the six new forts to be created. It seems the **Constante** was tasked with bringing munitions and money (20,000 pesos) from the commissary at Tampico, and the commander at Goliad had posted lookouts at Matagorda and Sabinito (a place name for the confluence of the Guadalupe and San Antonio Rivers at the head of current San Antonio Bay, near what was known as Rancho de Los Mosquitos, Los Mosquitos or Mesquite Landing), to assist in its arrival, and also to observe for the expected arrival of the **Oscar**. On 31-May-1830, apparently in an attempt to fulfill the provision for Mexican colonization of Texas, Terán also wrote to Andres Sobrevilla (commander at Laredo) and Martin de Leon (empresario of the Guadalupe colony – the boundaries of which can be seen in the Frontispiece and Figure 3), ordering them to assist a group of families traveling from Zacatecas to the Guadalupe Colony. He also wrote to Lucas Alamán about it, indicating this was in coordination with the Lavaca establishment [Terán 31-May-1830]. This was part of a larger plan by Terán to create settlements of 500 families each at Galveston (Bay area, probably Anahuac), Lavaca and Tenoxtitlán, which did not materialize [Morton 1945 p. 500]. Late in July, Terán wrote multiple letters when this group apparently was underway through Lampazos on its way to Laredo, indicating it consisted of 16 families and 31 "*presidiarios*" (prisoners/convicts/inmates), while ordering local authorities to assist in any way [Cosío 30-Jul-1830, frames 938-941b]. Later, Terán wrote to Ramón Músquiz, advising him that he'd ordered the commandants at these three forts to report when the goal of 40 families was reached, apparently since this goal had not been reached by that point [Terán 15-May-1831].

On 1-Jun-1830, Terán wrote a series of letters to Antonio Elosúa, Erasmo Seguin and Stephen F. Austin, informing them that troops from the 11th and 12th Permanent Battalions under the command of Captain Aniceto Arteaga were to embark from Matamoros on the sloop **General Bustamante** within two days to the mouth of the Lavaca, that a cavalry unit under Captain José Manuel Barberena (3rd Active Company of Tamaulipas) will march overland through Mier to La Bahía along with the Zacatecas families, and that Rafael Chowell was to be commissioner in building the Lavaca establishment. Chowell was trained as a mineralogist, and had accompanied Terán as a scientist in the 1828 boundary expedition. Attachments in the letters to Seguin include a copy of the specific instructions given to Chowell and Arteaga [Terán 1-Jun-1830, 5-Jun-1830].

The directives to Chowell include measurement of the latitude for the Lavaca anchorage (to help plot it on future navigation charts), to build a defensible structure for 100 soldiers at a flood-free site with fresh healthful water, questions about finding lime/clay/oyster shell and whether to build with brick or wood with the assistance of skilled laborers from Guadalupe or Austin's Colony, seek assistance from Martin de Leon about pastures and farming, to contact the Commissioner at Béxar (Erasmo Seguin) for funds if needed, and with knowledge that the state commander (Elosúa) was to be informed of all these plans.

The directives to Arteaga include taking 40 men from the 11th and 12th Permanent Battalions at Matamoros aboard the sloop ***General Bustamante*** to the anchorage at the Lavaca River in "*la Bahía de San Bernardo*", to create a military establishment <u>on the right bank</u> on the property of "*Colonia de Guadalupe*", with the purpose of keeping an eye on "*la Norte Americanos de Austin*" across the Lavaca, while protecting both colonies from the "*aggressions of the savages*", to be independent of the commander at La Bahia, but report his arrival to the regional commander (Antonio Elosúa at Béxar), and to otherwise communicate directly to Terán, to treat the North Americans with respect and contact Stephen F. Austin with honesty and enlightenment, to settle into the fort as soon as possible but stay at Guadalupe in the meantime, locate a cavalry company at La Bahia or Guadalupe, to construct using lime as directed by Rafael Chowell, and even advice about how to grow and process grain and also the importance of daily rifle maintenance, among other advice.

On the 5th of June, Terán wrote again to Elosúa, quoting messages he sent to Arteaga and Barberena, that Barberena was to locate his cavalry troop at La Bahía under Arteaga's overall command. Terán also wrote to Erasmo Seguin on the same day, requesting support for Barberena's trip [Terán 5-Jun-1830]. From all of these messages, we can see that Terán was choreographing a large coordinated effort to both colonize and establish military posts in this part of Texas, according to provisions of the new Law of 6-Apr-1830.

On the 6th of June, Cosío copied Elosúa with Terán's message of May 25th, and that he was additionally posting a lookout for the ***Constante*** at the Lavaca anchorage [Cosío 6-Jun-1830, frames 375-376b]. On the 9th, Elosúa wrote to Martin de Leon, requesting aid for unloading the ***Constante*** upon its arrival [Elosúa 9-Jun-1830], to which De Leon agreed [De Leon 18-Jun-1830].

The first word from Texas on the ***General Bustamante*** comes, strangely, from a dispatch written by Elosúa on 21-Jun-1830 from Béxar to Cosío at Goliad, acknowledging prior letters from Cosío (on the 17th) and Rafael Chowell (on the 15th) stating the vessel had been "*thrown to the beach*" on 11-Jun-1830, and orders all possible aid [Elosúa 21-Jun-1830, frames 831-832]. On the 28th, Elosúa writes to Terán, reporting the loss of the vessel, and including the Cosío letter of the 17th, indicating that the Chowell letter of the 15th was an enclosure [Elosúa 28-Jun-1830, frame 13]. Thus, the original copies of these initial reports (not found in the Béxar Archives) seem to have been sent off to Matamoros. No precise location is mentioned for the shipwreck.

One of the first things that Aniceto Arteaga did after arrival in Texas was to write to Elosúa, that he'd arrived at Guadalupe (Guadalupe Victoria, current Victoria, Texas) on 29-Jun-1830, and made an initial report of his forces [Arteaga 1-Jul-1830]. This report indicates that Arteaga had brought 1 officer and 35 infantrymen (total of 37 men), and their armament consisted of 33 muskets, 33 belts, 33 cartridge boxes and 66 gun flints – presumably the personnel and cargo brought and then rescued from the ***General Bustamante***. This was confirmed in a dispatch the next day from Cosío to Elosúa, based on a report from his men posted at the Lavaca anchorage that Arteaga and troops had disembarked at the bar of the Lavaca River on 26-Jun-1830 [Cosío 2-Jul-1830, frames 147-148]. So, perhaps the wreck lies somewhere near that Lavaca anchorage, and serves to illustrate the dangers of grounding in this shallow part of the bay.

The first muster reports or lists were made by a junior officer, Jóse María Castillo, on 3-Jul-1830 from Guadalupe, which were certified by Rafael Chowell and Aniceto Arteaga, indicating 24 members of the 12th Battalion and 12 members of the 11th were present [Castillo 3-Jun-1830]. Including Arteaga himself, this totaled 37 men, in agreement with the prior report.

In short order, Arteaga also wrote to Stephen F. Austin informing him of his new post, and that Terán suggested contact to request help locating laborers [Arteaga 6-Jul-1830]. Rafael Chowell also wrote to Stephen F. Austin from Guadalupe, establishing contact, indicating they had left Brazos Santiago (the harbor and anchorage near to Matamoros) on 7-Jun-1830 but *"you already will have known how badly fortune has treated us"* (the shipwreck?), and also requesting two subscriptions to the newspaper of San Felipe [Chowell 10-Jul-1830]. Austin promptly replied to Arteaga on 13-Jul-1830, with a diplomatic response but not committing to laborers; however, he offered to request such in the newspaper [Austin 13-Jul-1830]. Indeed, in the **Texas Gazette** issue of 22-Jul-1830, the following notice did appear. From this notice, one might infer that initial plans included construction of some wooden structures.

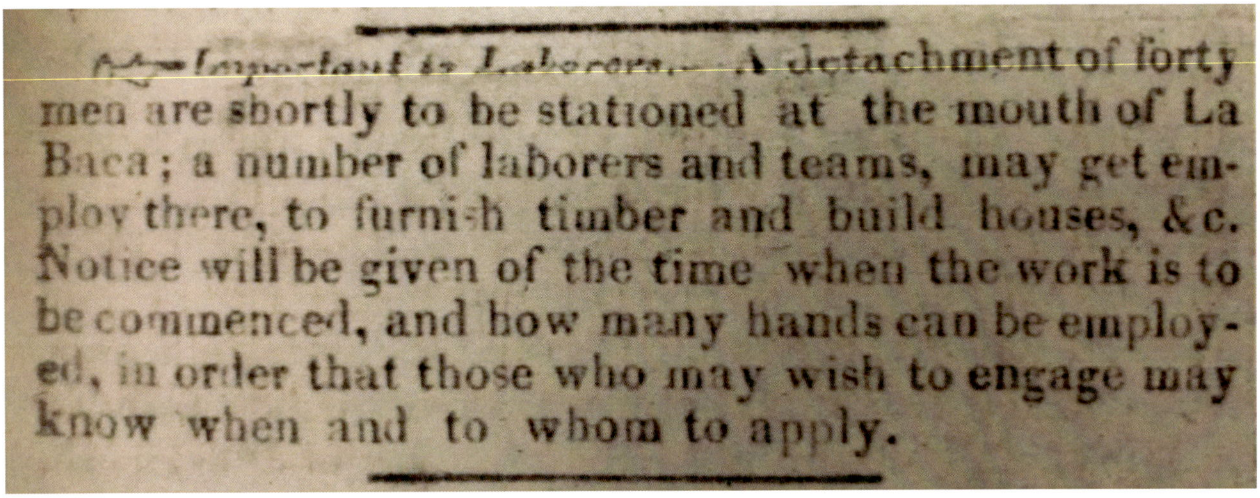

> *Important to Laborers.—* A detachment of forty men are shortly to be stationed at the mouth of La Baca; a number of laborers and teams, may get employ there, to furnish timber and build houses, &c. Notice will be given of the time when the work is to be commenced, and how many hands can be employed, in order that those who may wish to engage may know when and to whom to apply.

Figure 20: Notice in Texas Gazette, 22-Jul-1830, Page 2, Column 1

Later in July, Arteaga received a letter from Elosúa, passing on a message from George Fisher on the Brazos, suggesting to aid Fisher or use of the schooner **Cañon** [Elosúa 21-Jul-1830], which had been seized for importing contraband tobacco, so as to recoup the large imposed fine.

Terán wrote to Elosúa again informing him that supplies were coming to the port of Matagorda from New Orleans for troops in Texas, and these should be unloaded through the new Lavaca post, with money to be directed to Nacogdoches via Tenoxtitlán [Terán 17-Jul-1830]. Indeed, two schooners (**Pomona** and **Rover**) arrived on 21-Jul-1830 at Aranzazu from New Orleans, laden with supplies for troops in the Department of Texas, consigned again to Salvador Flores [Cosío 30-Jul-1830, frames 942-942 & 944-945], now including provisions for the new Terán forts. Perhaps these were the long-overdue supplies which the **Oscar** tried to bring earlier that year.

Beginning about July 1st, a practice began of making formal military reports to Antonio Elosúa, on or about the 1st of each month that reported the status of the detachment as of that point in time. To some degree, Elosúa clarified the information he wished to receive, including also reports on ship arrivals in the area [Elosúa 19-Aug-1830]. These reports took various forms, but continued on a regular basis until the Lavaca post was abandoned in the summer of 1832. One type was usually issued by the Captain of the detachment, and is often referred to as a "monthly military report" in the Béxar Archives Calendar, but was typically a high-level one-page tabular summary of the headcount and available armaments, and usually was forwarded to Elosúa with a cover letter dated a few days later into the month. This type of report is illustrated in Figure 21 below for the month of Aug-1830 [Arteaga 1-Aug-1830]. In this example, the 11th Permanent Battalion is represented with no officers, 1 Sergeant, 1 Corporal and 10 soldiers, the 12th Permanent Battalion by 1 Lieutenant, 1 Sergeant, 2 buglers and 21 soldiers. In addition, there was the overall commander (Arteaga), making the total headcount of 38 present in Guadalupe at this early stage, obviously reflecting the troops that arrived on the *General Bustamante*.

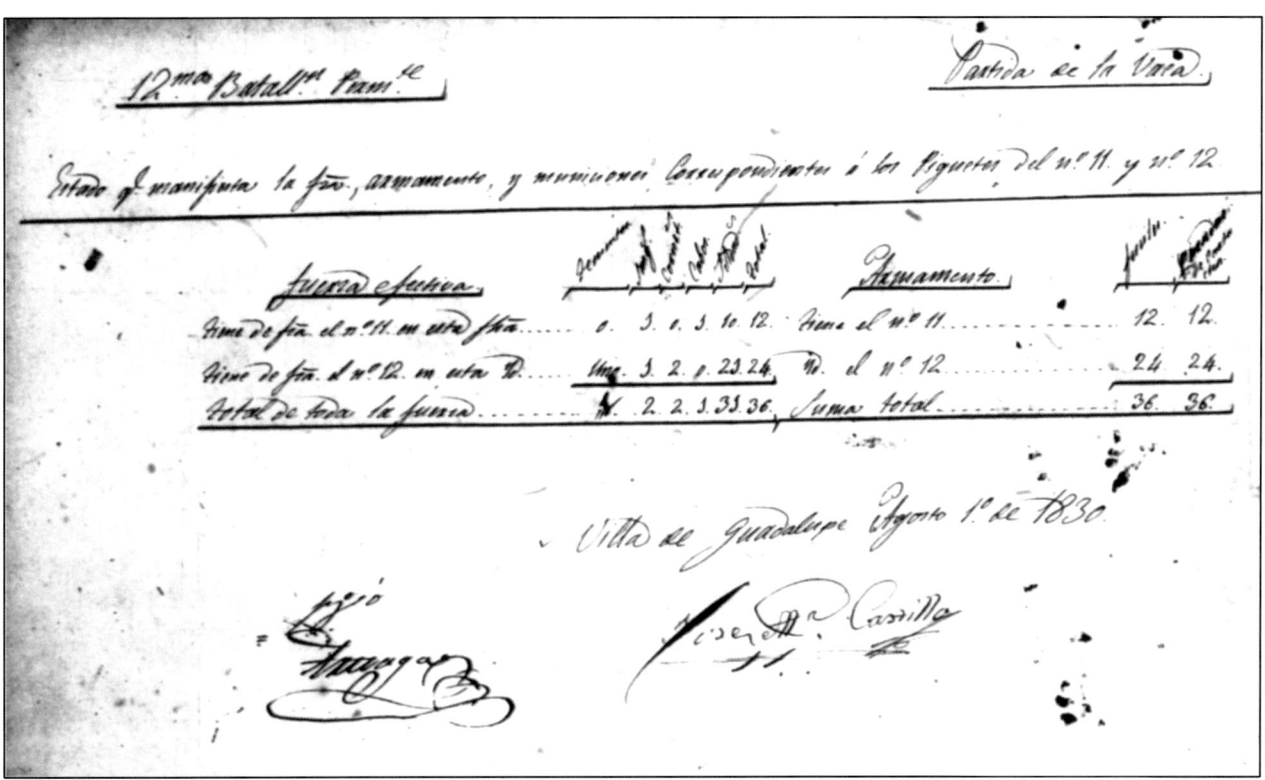

Figure 21: Monthly "summary" report by Arteaga and Castillo from Guadalupe dated 1-Aug-1830

As Arteaga's command grew, this type of report grew in complexity. Just two months later, the report for Oct-1830 included additional information, sometimes including side notes, as shown in Figure 22 [Arteaga 1-Oct-1830]. In this example, Arteaga lists himself as a Captain from the 3rd Permanent Battalion, 2 Lieutenants, 2 Sub-lieutenants (also sometimes called "Alférez"), 6 Sergeants, 2 Buglers, 6 Corporals and 103 soldiers. Armaments include 117 muskets (with bayonets), 98 belts, 19 horses and 1080 cartridge boxes. A note at the bottom indicates that Reverend Friar Miguel Muro was appointed by Terán to serve as chaplain in concert with the garrison at La Bahía.

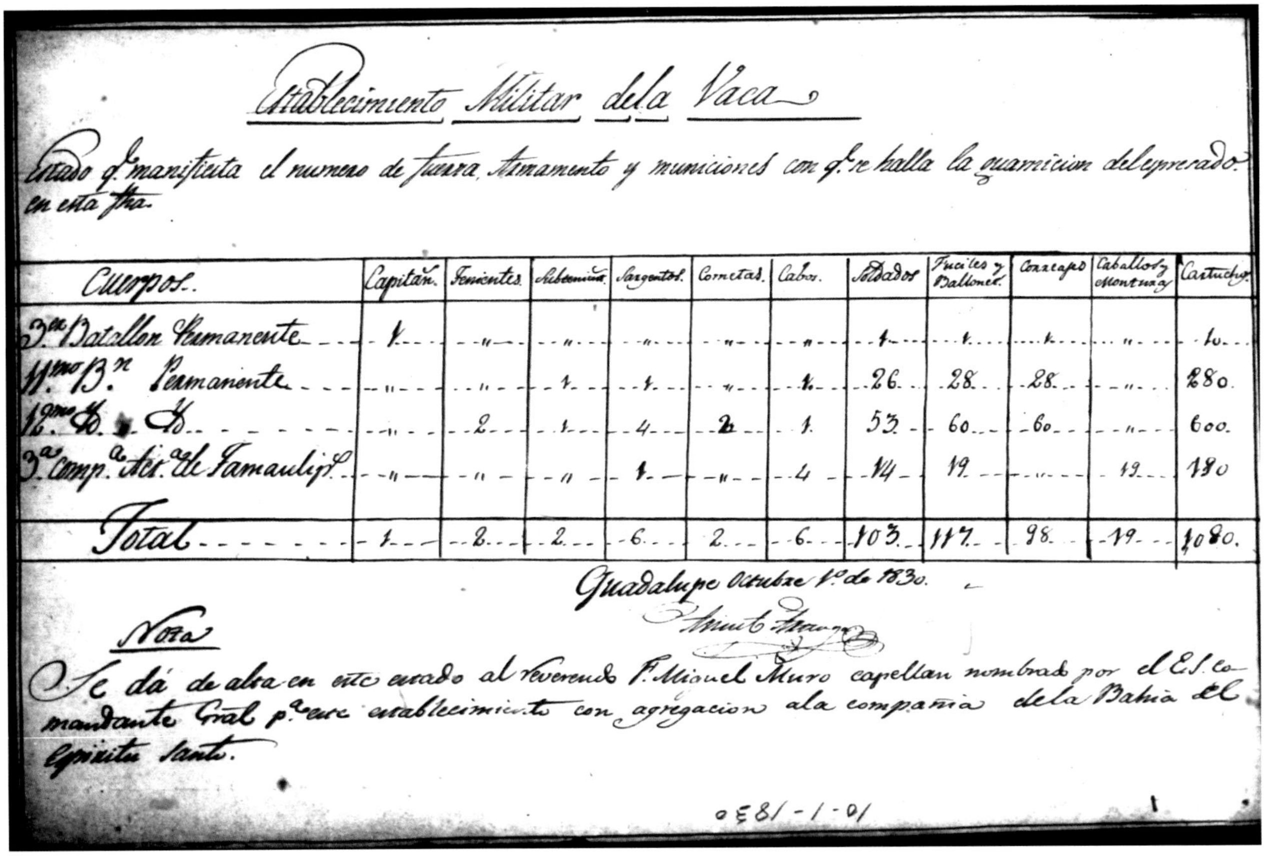

Establecimiento Militar dela Vaca

Estado q.e manifiesta el numero de fuerza, Armamento y municiones con q.e se halla la guarnicion del expresado en esta fha.

Cuerpos.	Capitan.	Tenientes.	Subtenientes.	Sargentos.	Cornetas.	Cabos.	Soldados.	Fusiles y Ballonetas.	Cananas	Caballos y Monturas	Cartuchos
3.er Batallon Permanente	1	"	"	"	"	"	1	1	1	"	10
11.mo B.n Permanente	"	"	1	1	"	1	26	28	28	"	280
12.mo Id. id.	"	2	1	4	2	1	53	60	60	"	600
3.a Comp.a Act.a de Tamaulip.s	"	"	"	1	"	4	14	19	"	19	190
Total	1	2	2	6	2	6	103	117	98	19	1080

Guadalupe Octubre 1.o de 1830.

Aniceto Arteaga

Nota

Se dá de alta en este estado al Reverendo P. Miguel Muro capellan nombrado por el E.S. Comandante Gral p.a este establecimiento con agregacion ala compañia dela Bahia del Espiritu Santo.

Figure 22: Monthly "summary" report by Aniceto Arteaga from Guadalupe Victoria, 1-Oct-1830

Another type of report, usually described in the Béxar Archives Calendar as a "commissary review report" or sometimes "roster report", is what might also be called a "muster report" and was often written by a junior officer but certified by Arteaga and Chowell, shown below in Figure 23 [Castillo 3-Aug-1830a]. These reports often included explanatory side notes, such as mention of deaths or desertions. In this example, we see the roster of the 11th Battalion at Guadalupe, including the names for their Sergeant (Marselo Leon), Corporal (José Maria Ramos) and 10 soldiers. A similar report is found for the 12th Battalion, listing a total of 41 enlisted personnel and 1 officer, but with 17 absent [Castillo 3-Aug-1830b]. Compared with the July report, we see the arrival of 12 members of the 11th and some new members of the 12th – perhaps these were those brought on the ***Constante*** (since it first detoured to Tampico), which had arrived at some point during July. There is also no indication yet of convicts.

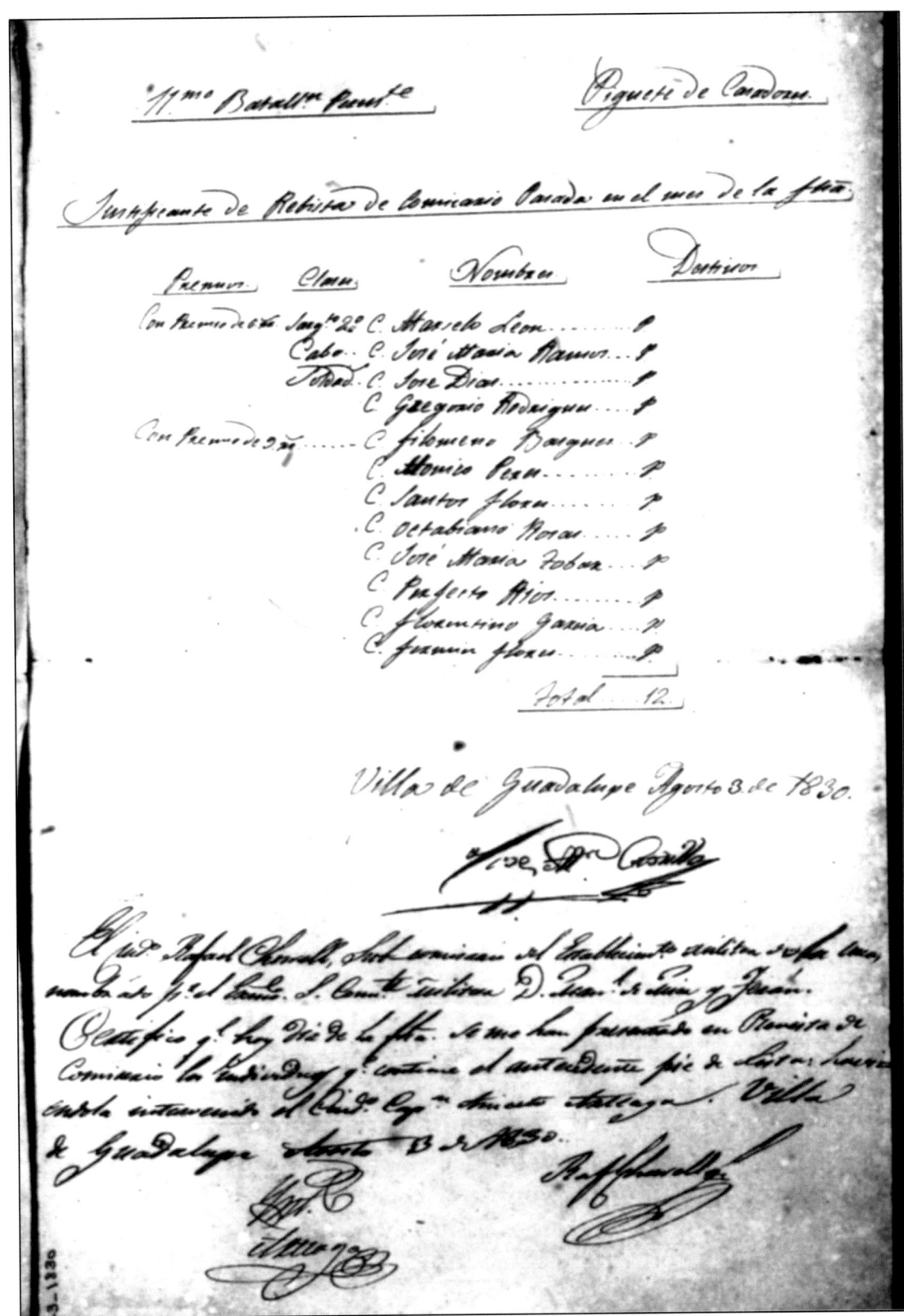

Figure 23: Monthly "muster" report by Castillo from Guadalupe dated 3-Aug-1830

One might notice that, near the bottom side of Figure 22, there is some handwritten text in which Rafael Chowell certifies that Aniceto Arteaga was present for the commissary review. However, for the first few months, separate documents or pages were also used to certify the presence of certain individuals, especially Aniceto Arteaga. An example of such for this same review (3-Aug-1830) is illustrated in Figure 24 below [Chowell 3-Aug-1830]. Please note that the actual signature of Rafael Chowell in Figures 18 and 19 seem to use the spelling "Chowell" for his surname, so that is the spelling adopted here, although other spellings (Chovell, Chovel, Choval) are sometimes observed.

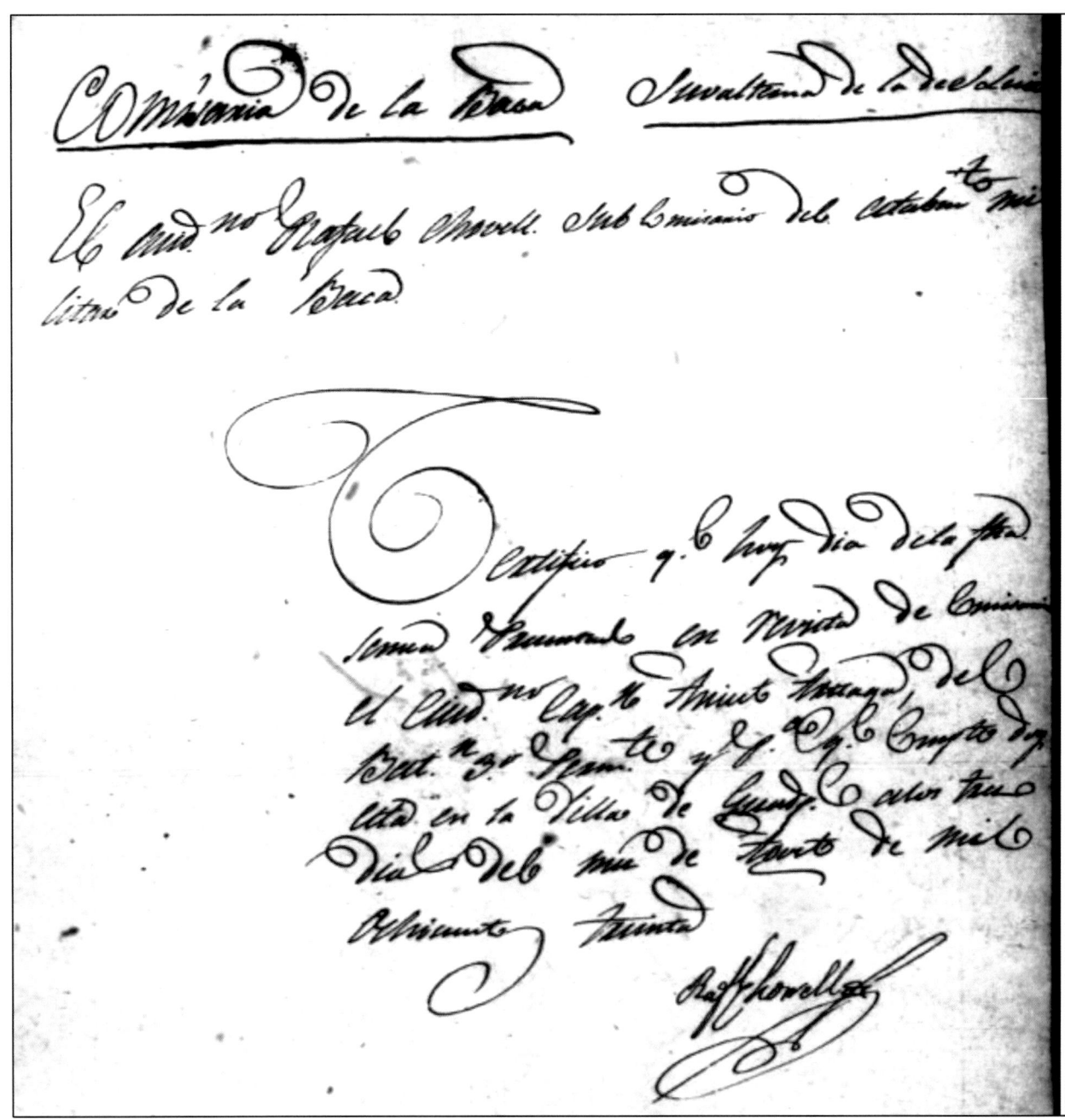

Figure 24: Certificate of attendance for Aniceto Arteaga at commissary review, by Rafael Chowell

Over at La Bahía (Goliad), Captain Barberena had apparently arrived before August, and made a similar report for his command (3rd Active Company of Tamaulipas), originally also a tabular summary report, as shown in Figure 25 below [Barberena 1-Aug-1830]. In this report, we see that many of his cavalry troops are deployed outside of Goliad, including 15 stationed at Guadalupe (red outlined area). Counting these 15 cavalrymen (and the absent infantrymen), the total headcount assigned to the Lavaca detachment at Guadalupe can be estimated as 70 by early August.

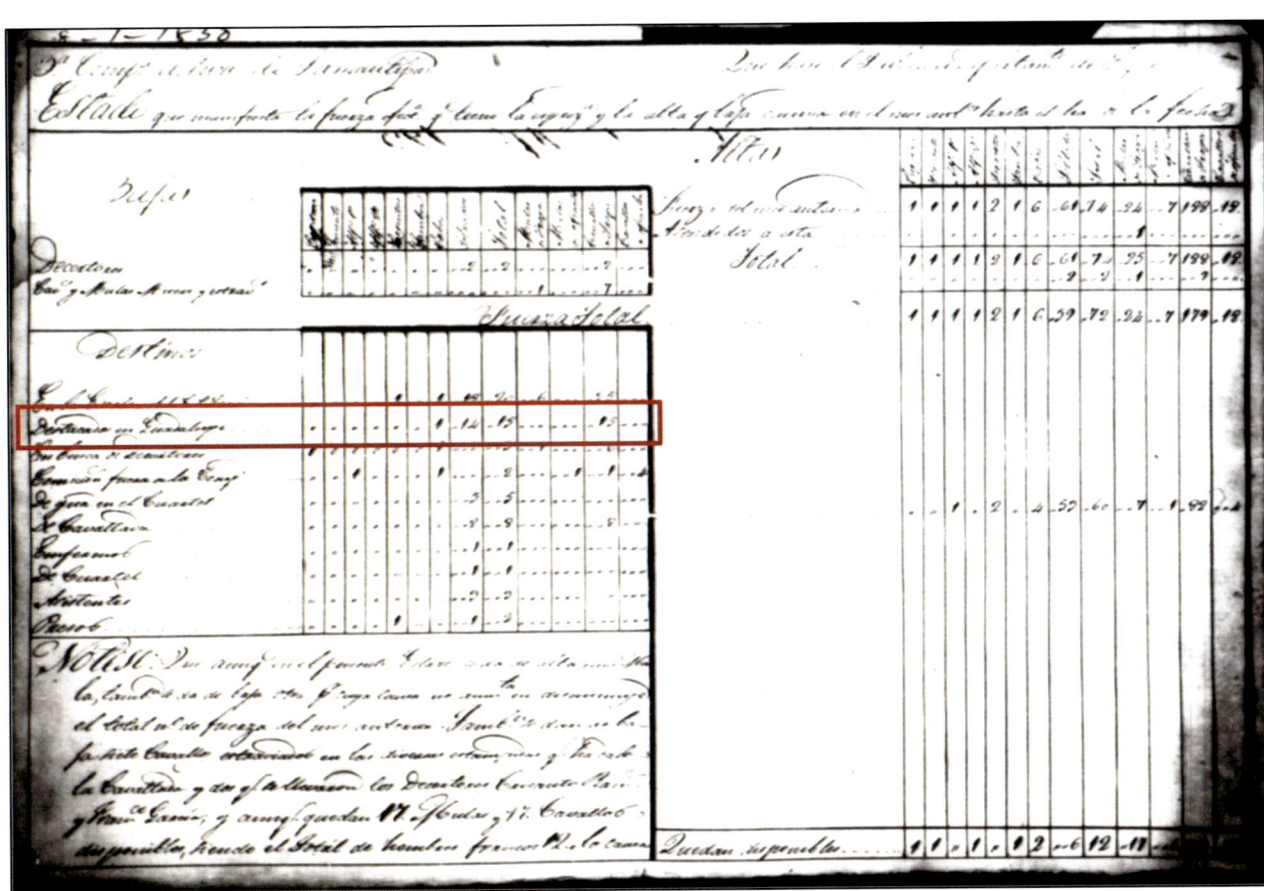

Figure 25: Monthly "summary" report by Barberena from Goliad dated 3-Aug-1830

Later in August, a second trip of the **Constante** brought 58 additional soldiers and 2 Sub-lieutenants (Miguel Zenon Trujillo and Manuel Osores), which arrived at Aransas Point on 12-Aug-1830, although 14 men from the 12th deserted after arrival [Arteaga 19-Aug-1830]. On this trip, 13,500 pesos were brought to Rafael Chowell, of which he forwarded 7656 pesos and 3 reales to Erasmo Seguin at Béxar [Chowell 19-Aug-1830]. It must have also been sometime in August when the first convicts arrived at Guadalupe, as 34 "presidiarios" are first listed in their own roster report at the beginning of September [Trujillo 3-Sep-1830], shown in Figure 26. Also, by this time, an additional 20 cavalry were also posted to Guadalupe from the 3rd Co. at Goliad, bringing the total headcount to 141 at Guadalupe.

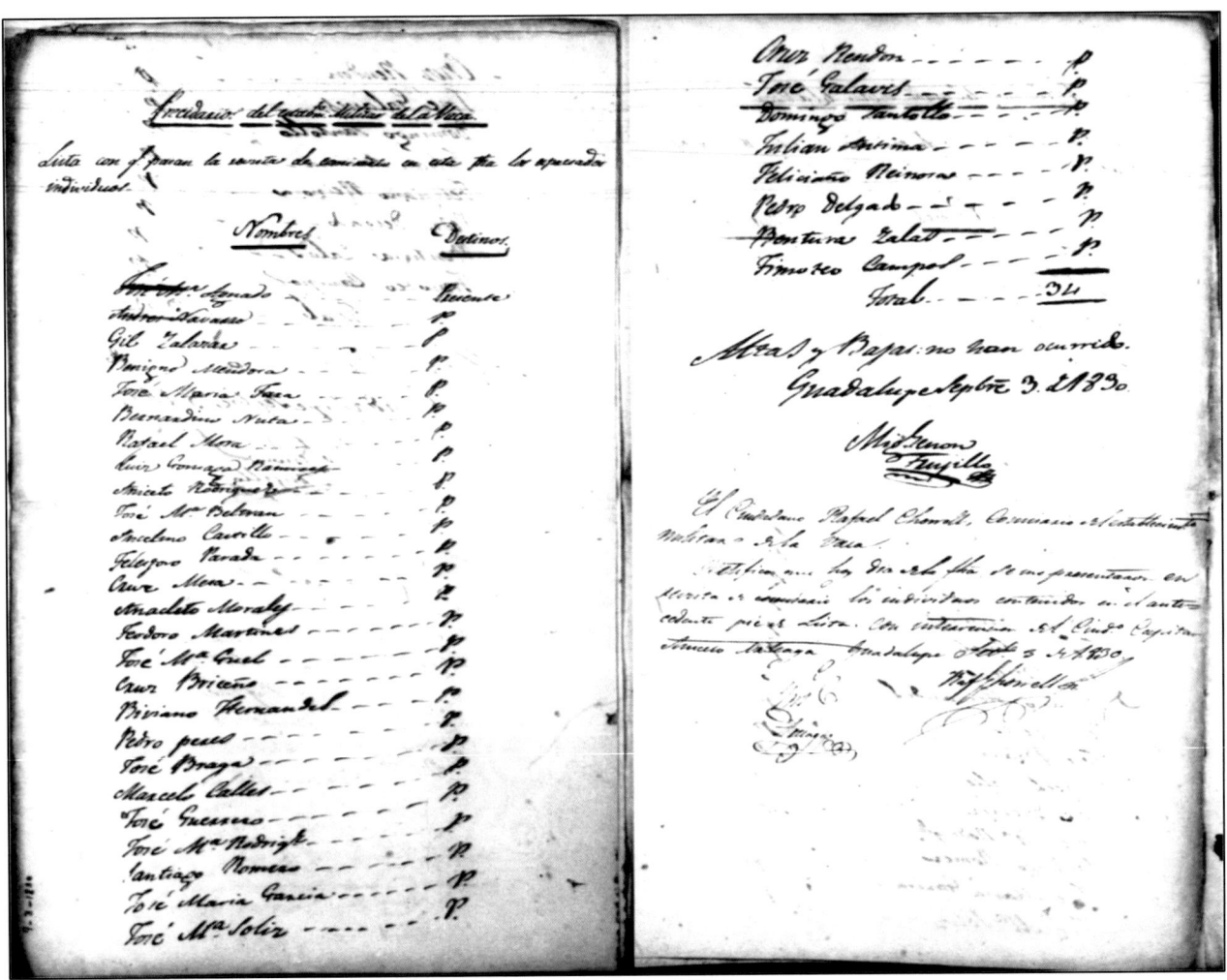

Figure 26: Commissary Review Report for Convicts by Trujillo at Guadalupe, 3-Sep-1830

Interestingly, the monthly salary of the various positions is revealed in another type of report, usually referred to as a "budget report" authored by the commissioner or unit paymaster. Most especially in those for the 3rd Active Company, it can be seen that they itemized each role, an example from Feb-1831 is shown in Figure 27 below – Captain (125 pesos), Lieutenant (66 pesos, 5 reales, 4 granos), First Alferez (50 pesos), Second Alferez (41 pesos, 5 reales, 4 granos), Sergeant (30 pesos), Drummer (12 pesos), Corporal (25 pesos) and Soldier (20 pesos). Convicts were to receive a stipend of 8 pesos per month to a common account, be dressed in coarse brown or green cloth to distinguish them from the soldiers, provided with sufficient rations for their families, yet allowed a ration of brandy or whiskey [Terán 19-Sep-1830, Chowell 12-Oct-1830].

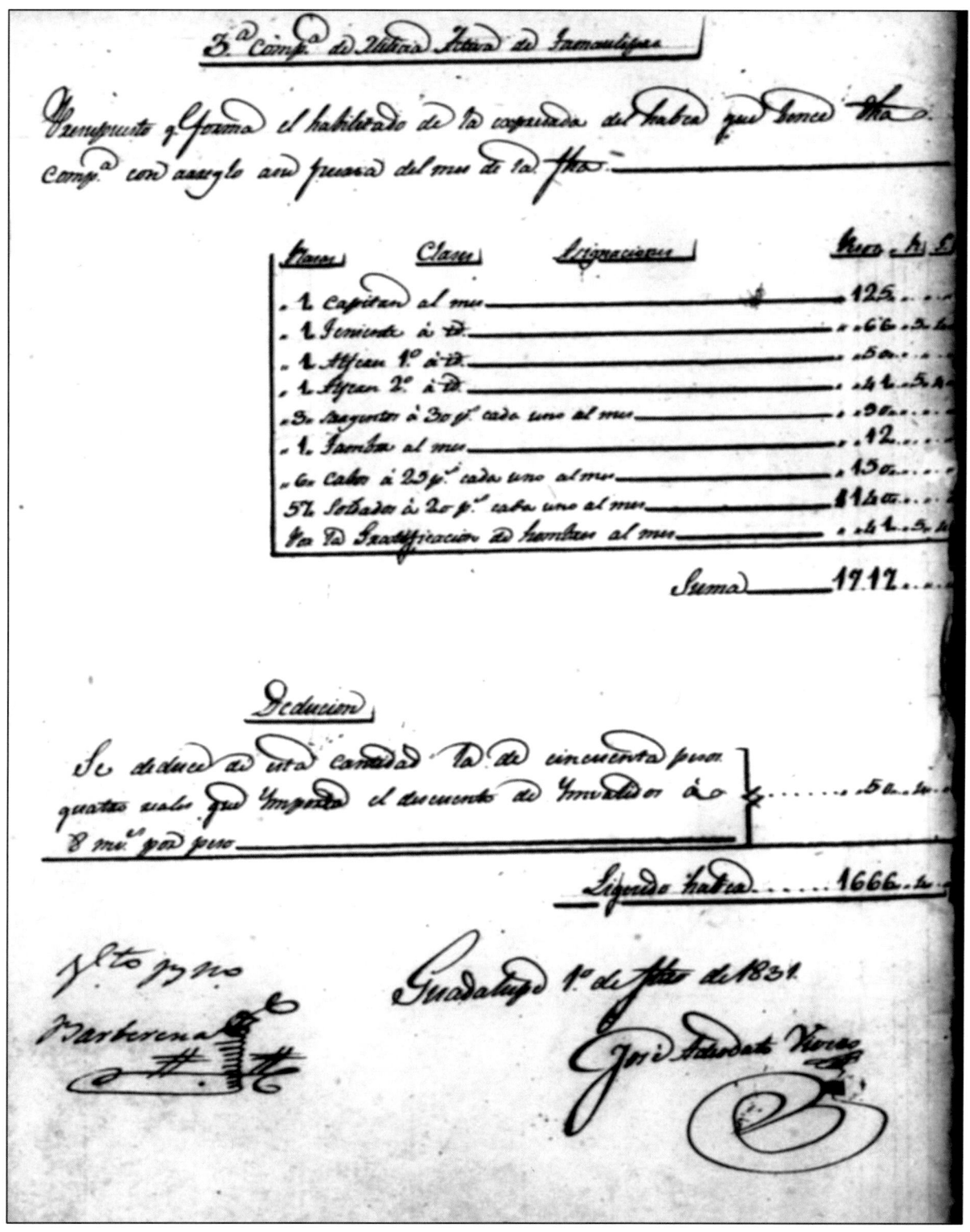

Figure 27: Budget Report for 3rd Active Company by José Adeodato Vivero dated 1-Feb-1831

Another voyage of the **Constante** arrived at "*Garcitas*" (presumably the anchorage at the head of Lavaca Bay near the mouth of Garcitas Creek) on 22-Sep-1830, bringing 75 recruits (32 from 11th, 43 from 12th) and 6000 pesos [Arteaga 12-Oct-1830]. These personnel numbers do not show up in the subsequent reports for the Lavaca detachment, so many of these soldiers ended up as reinforcements for other posts. A letter from Terán indicates these troops were new recruits from Nuevo Leon, gathered at Matamoros, and then sent on to Lavaca. They were intended for the Nacogdoches post commanded by José de las Piedras, and he sent Lt. Juan Perez de Arze from Nacogdoches to retrieve them; in the meantime they were housed at Lavaca and supplied by Chowell [De Las Piedras 2-Aug-1830]. The money, though, was intended for Rafael Chowell and the Lavaca post [Terán 15-Sep-1830]. This reference also indicates that Terán was already encountering some delays in obtaining the full budgeted funding from the Mexican Treasury for his forts in Texas, estimated at 8000 pesos annually per post.

The monthly reports for Jan-1831 showed that 34 convicts were present (now at Barranco Colorado), along with 24 soldiers of the 12th Permanent Battalion. Each group had also suffered one death in the prior month. An additional 35 soldiers were posted at Guadalupe or other places, some perhaps on leave [Trujillo 3-Jan-1831, Castillo 3-Jan-1831].

Since the various military reports are scattered throughout the Béxar Archives over about 2 years, and often separated from each other and their cover letters, it would not be especially productive to separately discuss each and every one. The examples mentioned above in Figures 21 to 27, and all others located to date, are indexed in the following table.

Date	Cover Letter	11th Battalion	12th Battalion	3rd Co. (Cavalry)	Summary	Attendance Certified	Presidiarios
1-Jul-1830	132:115-116a	132:232-234	132:229-231		132:116b		
1-Aug-1830	133:24, 458-459, 796	133:141-142	133:143-145	133:10-13	133:24	133:146-147	
1-Sep-1830	134:279, 297-298	134:96-98	134:81-88	134:5, 261-262	134:33	134:70, 80	134:91-95, 89-90
1-Oct-1830	135:351	134:990-992	134:984-987	134:887-890, 896	134:934	134:980-983	134: 988-989
1-Nov-1830	135:959, 965	135:966-968a	135:968b-971	135:904-907, 942-943; 136:37-39	136:61-62	135:960-962	135:972-974, 940-941
1-Dec-1830		136:781-783	136:774-777b	136:657-660, 664-666	136:702-704	136:784-787	136:778-780
1-Jan-1831	137:786, 795	137:574-576	137:591-594	137:482-489	137:494-495	137:577-580	137:588-590
1-Feb-1831	138:591	138:592-594a	138:594b-597b	138:435-442	138:492-493	138:443, 600b-602	138:598-600a
1-Mar-1831	139:280			139:64-72	139:63		
1-Apr-1831	139:835			139:827-833	139: 836		
1-May-1831	140:177			140:241, 518-523			
1-Jun-1831	141:495-7,683 142:159-61			141:448, 508-510, 663-664	142:112		141:498-499
1-Jul-1831	143:182			143-603			
1-Aug-1831	152:205-206, 143:261			143:379, 380-383, 514-515	143:385		
1-Sep-1831	144:255, 301, 303			144:185-188	144:178		144:179-180
1-Oct-1831	153:773			145:28-29, 98-100	145:27		153:765-767
1-Nov-1831	153:794			145:702-706, 850	145:694-695		153:791-793
1-Dec-1831	154:253				146:365-366		146:363-364
1-Jan-1832	147:219				147:17-18		147:15-16, 217
1-Feb-1832					147:800-801		147:802-803
1-Mar-1832	148:763				148:337-338		
1-Apr-1832	149:282-284,335				149:76		149:77-78
1-May-1832	150:356				149:719-720		
1-Jun-1832	150:736, 740				150:458		150:433-434
1-Jul-1832	151:548, 637-638						

Table 1: Index to all known Lavaca monthly military reports found in the Béxar Archives (Roll:Frames)

Table 1 indicates that the unit "muster" reports (and their certification letters) were used only through Feb-1831; however, the "summary" and "prisoner" reports were used for the entire period. Empty cells indicate that some reports are not found, although the majority are available in the Béxar Archives. The pink-tinted cells were addressed from Barranco Colorado, and the uncolored cells from Guadalupe. Several of the first reports of the 3rd Co. were from Goliad, in the green-tinted cells.

A census of sorts was made based on these reports, and a graph made of the military population over time (see Figures 28 and 29 below). These graphs reveal major shifts, which are discussed after the graphs, along with other major developments.

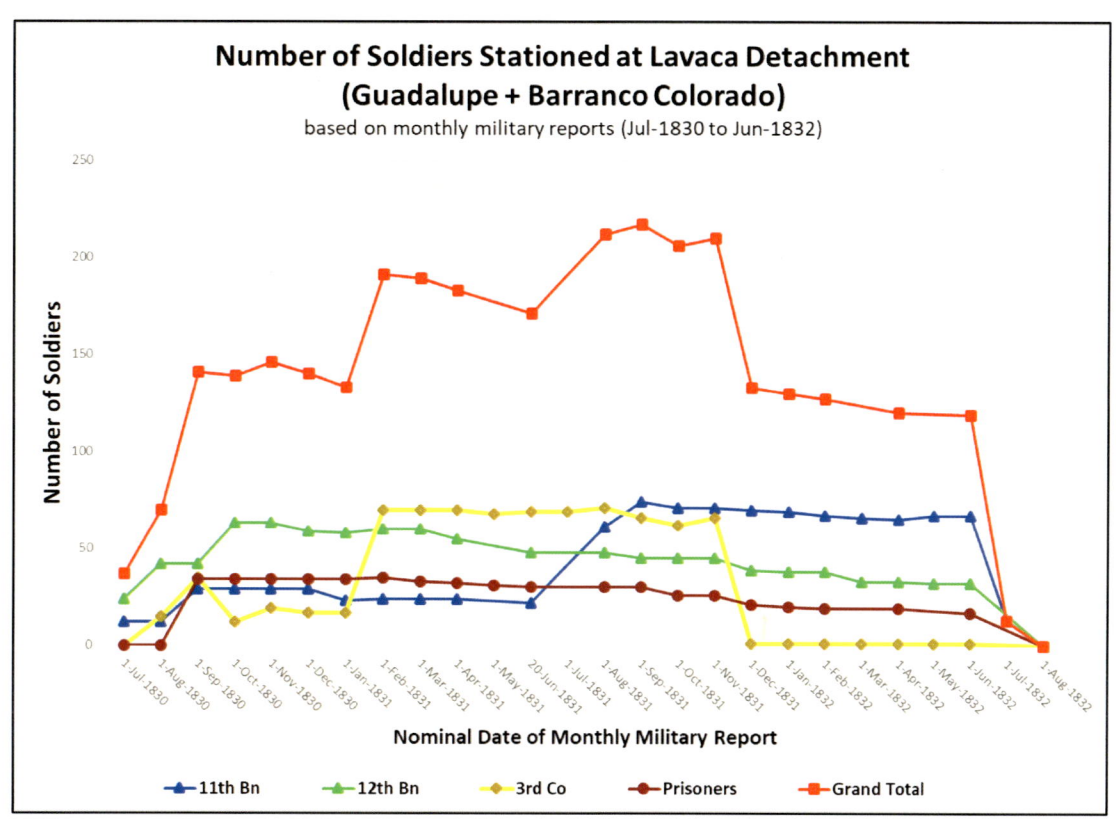

Figure 28: Military Population of Lavaca Detachment (with Grand Total)

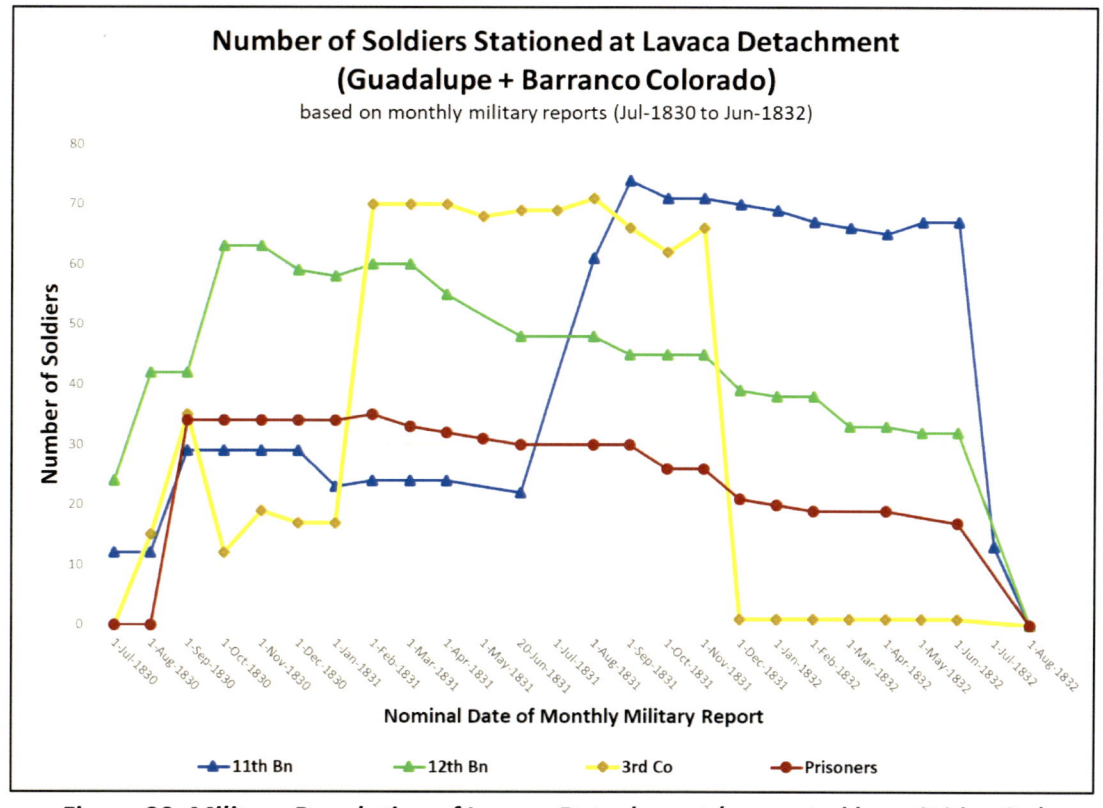

Figure 29: Military Population of Lavaca Detachment (separated by unit identity)

From the discussion above, and the graphs in Figures 28 and 29, it can be seen that there was an initial growth period where several deliveries of soldiers and convict laborers brought the Lavaca detachment's size up to about 140-150 personnel for a period of Sep-1830 to Jan-1831. Initially, all of the reports were from Guadalupe, indicating that Arteaga used the existing town of Guadalupe Victoria as a base, perhaps while investigating the area and selecting a specific site for Barranco Colorado. However, beginning with the reports of 1-Oct-1830, the location of the 12th Battalion and the convict group switches to Barranco Colorado, perhaps indicating the convicts were put to work building a brick kiln and initial structures, and the 12th Battalion were their guards. One document suggests this was to include "... *manufacture of barracks*" [Chowell 19-Aug-1830]. This arrangement continued through early January of 1831.

In Aug-1830, it can also be seen that Terán was also planning for the religious welfare of the soldiers, prisoners and settlers, since he wrote to one Father Miguel Muro, soliciting him to become the chaplain at Lavaca with an annual salary of 500 pesos and an inducement to pacify the local Karankawa and Jaraname tribes through settlement and work at the new establishment [Terán 21-Aug-1830]. Muro's assignment apparently came to pass as indicated in Figure 22. So, again, we see more evidence of Terán's master plan. Indeed, within later monthly military reports, a single "Capellán" (chaplain) is reported present at Barranco Colorado from Feb-1831 to Jun-1832.

During this period, Terán apparently felt the initial directives to Chowell and Arteaga were not sufficient, and he created a model and plan for construction of a fort, which was sent to them in the Fall of 1830. In a letter to Chowell accompanying these items, he wrote "*In a wooden box I am sending you the model and plan which is to serve for the construction of the fort. Please note that the thickness of the walls has not been drawn to scale, and they should have the same thickness as any ordinary house, in proportion to the material with which they are constructed. Around this building, for a distance of 400 varas (about 1100 feet) no building shall be permitted, for it is to serve as the citadel in the settlement which is to be formed in the course of time. Also you should try to clear the ground of brush and any other objects which might tend to limit the effectiveness of firearms.*" [Terán 12-Oct-1830]. Terán also instructed Chowell to take down the details, and then promptly forward the model and plan on to Antonio Elosúa for use in building Fort Tenoxtitlán, which Chowell did the next month [Chowell 8-Nov-1830 & 17-Nov-1830]. Terán had also notified Elosúa (who notified the commander at Tenoxtitlán, José Francisco Ruiz) of the plan. Ruiz responded by requesting it be sent to him, and Elosúa did so in early December [Terán 20-Sep-1830]. Interestingly, Ruiz sent the box south again with Ensign Santiago Navayra to Samuel May Williams (at San Felipe), asking him to translate the plan into English (so that Anglo-American laborers could read it) and also the ensign was "*to look for a man capable of building the said house*" since Ruiz wrote "*I find myself without the necessary knowledge for building fortifications*" [Ruiz 26-Dec-1830]. A footnote in this reference notes with irony "*... that apparently Ruiz was going to employ Anglo Americans to build this fort whose purpose was to keep Anglo Americans out of Texas.*". Although the model and plan have apparently not survived, the important point is that Barranco Colorado and Fort Tenoxtitlán may have the same design for any main structure. Jack Jackson also felt that this design was to be used for Fort Lipantitlán [Jackson et al 2006 pp. 63-65]. Thus, if archival or archaeological evidence is found for any one of these, it might be concluded the others resembled such a design as well.

The graphs then reveal a major increase in the number of the 3rd Active Company beginning with the Feb-1831 report, and the location of Barberena's reports switches from Goliad to Guadalupe, based on an order to move there from Terán [Terán 30-Oct-1830]. Dispatches in the Béxar Archives tell us that Arteaga moved his command to Barranco Colorado on 7-Jan-1831, and that Barberena moved from Goliad to Guadalupe to take command there [Arteaga 12-Jan-1831], in what was Arteaga's first letter written from Barranco Colorado. Now counted as part of the "Lavaca detachment", this cavalry unit increases headcount to almost 200.

As the summer of 1831 arrived, the Lavaca detachment was afflicted with an epidemic of fevers and chills. The first report was from Guadalupe, where Captain Barberena wrote that he was suffering an illness caught on his way from Tampico, and that he could not travel to Goliad, where Colonel Elosúa was making a visit [Barberena 12-Jun-1831]. Ten days later, Arteaga first reports that fevers are causing problems at Barranco Colorado [Arteaga 22-Jun-1831]. A week after that, Arteaga reported he had so many *"patients and convalescents"* that he could not cover essential services such as guarding a schooner (the **Hetta**?). He also wrote requesting quinine and *"the necessaries to contain this evil somewhere."* By the end of July, Arteaga reported that the fevers had continued, and he had 23 men of the 11th (virtually the entire complement), 18 men of the 12th and 16 prisoners (about half of each group) that were gravely ill. He also requested relief from the troops at Goliad, needs for building a hospital, and that he might move to Guadalupe. Elosúa responded by asking the Goliad commander to supply infantry troops of the 11th Battalion there to relieve Arteaga, and acknowledges the difficulty to occupy the confiscated **Hetta**. Work had also stopped on the fort. By the end of August, Arteaga requested additional medicines [Arteaga 28-Jun-1831]. The symptoms suggest the illness was probably malaria and/or yellow fever (spread by mosquitos), diseases which were not uncommon during this period in coastal Texas.

A major influx of troops from the 11th Battalion occurred in the period of Jul-1831. Apparently, a fresh group of about 75 soldiers and officers (and their provisions) of the 4th Company of the 11th Permanent Battalion of Tamaulipas was sent to relieve those few posted at Barranco Colorado, under orders from José Mariano Guerra [Guerra 10-Jun-1831], who had been placed in temporary command at Matamoros while Terán was away [Terán 16-May-1831]. Subsequent military reports at Lavaca show about this same number of troops of the 11th at Barranco Colorado (with almost identical makeup), and the previous report (20-Jun-1831) showed 22 members of the 11th. So, it is assumed that the net increase of about 50 troops of the 11th observed after this point was due to an exchange of troops of that Battalion, pushing headcount to above 200 (the figure mentioned by Linn). As already stated, it also appears the illness experienced at Barranco Colorado helped expedite the move of these reinforcements.

It was also in the summer of 1831 that the seizure of the schooner **Hetta/Hesta** occurred, alluded to in the aforementioned account by John S. Menefee. At Barranco Colorado, first notice of its presence was apparently the arrival of a (small?) boat loaded with goods that arrived there on or about 4-Jul-1831, after which the schooner itself was confiscated in Lavaca Bay due to its reputation for handling clandestine cargo. The ship was first guarded by Lieut. José María Castillo, a Sergeant and 12 soldiers (perhaps the very ones in Menefee's story). Local authorities such as Juan José Hernandez, Bonifacio Galan (Commissioner at Goliad) and Rafael Manchola (Alcalde of Goliad) adjudicated and confirmed the seizure [Arteaga 6-Jul-1831]. Later reports revealed its full cargo, as shown below in Figure 30. These included many crates, each marked for its intended recipient, and included Flour, Biscuits, Beans, Fish,

Sugar, Limes, Coffee, Coffee Grinders, Pepper, Butter, Brandy, Soap, Tobacco, Nails, Knockers, Plowshares, Iron, Steel, Knives, Candlesticks, Crystal, Plates, Pencils, Brushes, Padlocks, Spoons, Pots and many other items. The crew of the schooner were held until 16-Aug-1831, and were thought to have contributed to the disease afflicting the garrison. The ship furnishings were also confiscated, and the ship was left bare and abandoned in the bay, perhaps because Arteaga had too few healthy troops to compose a prize crew.

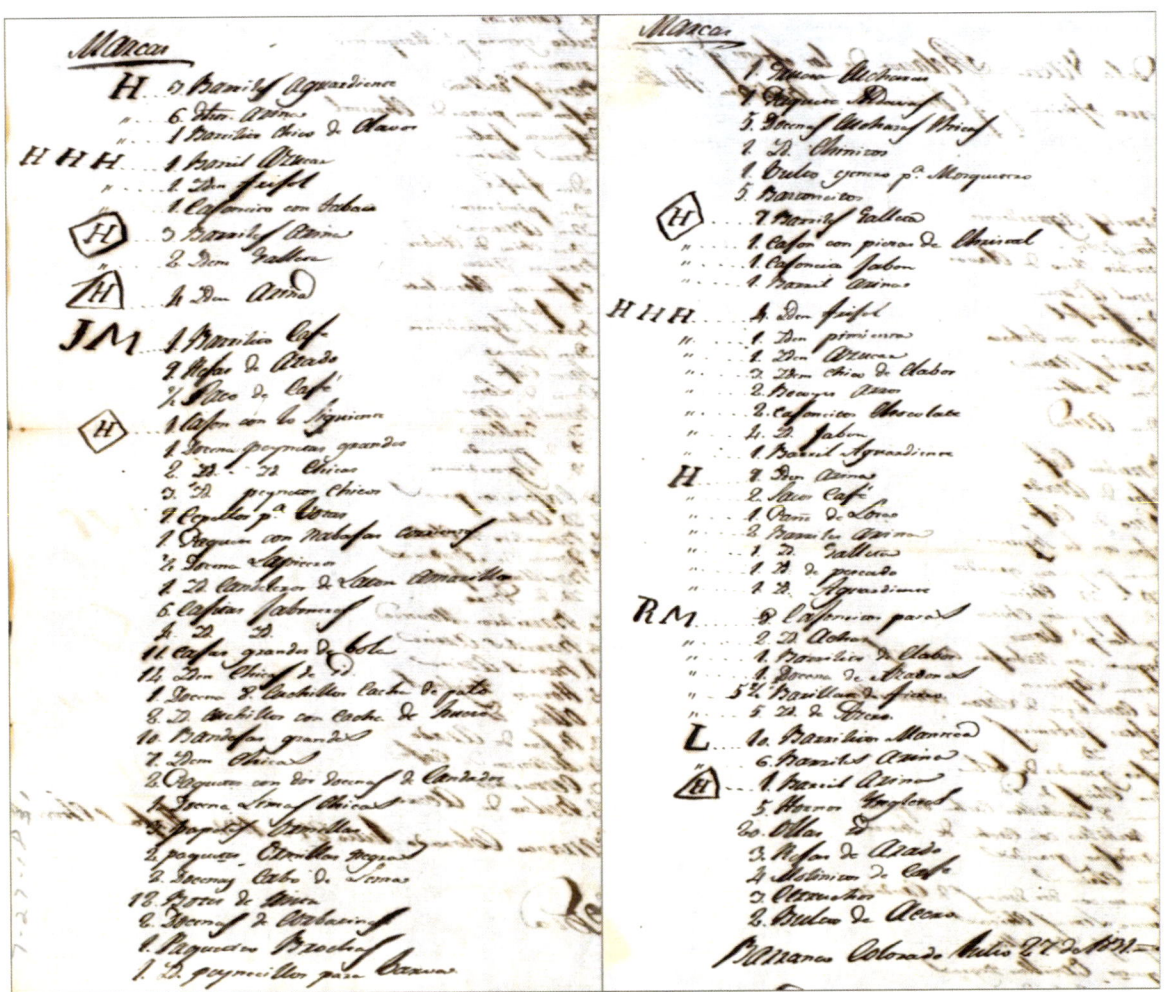

Figure 30: Goods confiscated from the schooner Hetta

After about 10 months in Guadalupe, the 3rd Active Company was withdrawn on 9-Nov-1831 [Barberena 9-Nov-1831] based on an order from Terán to both Elosúa and Barberena for the entire unit to be gathered together and then travel to Matamoros [Terán 24-Sep-1831]. In response, Elosúa ordered Cosío to supply a small group of cavalry (1 Corporal and 4 soldiers) from Goliad to relieve the 3rd Company at Guadalupe (presumably to maintain the mail service), which arrived there on 31-Oct-1831 [Elosúa 21-Oct-1831].

Beginning in late 1831, the dispatches in the Béxar Archives show pleas for provisions and supplies, which become critical by the summer of 1832. The first report is a complaint that the few cavalry troops supplied from Goliad in replacement of the 3rd Company were poorly mounted and provisioned [Arteaga

11-Dec-1831]. By Apr-1832, Arteaga complained that Barranco Colorado had not been supplied since January [Arteaga 11-Apr-1832]. Elosúa even asked Erasmo Seguin about a loan to provide provisions [Seguin 16-Apr-1832], and wrote to both Seguin and Ramón Músquiz to summon aid in the form of money, meat and other food from Goliad and Guadalupe Victoria [Elosúa 16-Apr-1832a]. The stress of this situation apparently caused one of the young lieutenants who had authored many muster reports for Barranco Colorado, José María Castillo, to be declared insane [Elosúa 16-Apr-1832b].

The schooner *Cañon* and its captain (Edward Pettit) had played a role in an unusual controversy about tobacco smuggling at the Brazos in the summer of 1830, when seized by George Fisher in his abortive attempts to establish the *"Aduana Maritima de Galvezton"* at that location. However, the vessel then appears absent from the records in the Béxar Archives until early 1832 at the port of Matagorda, when it is again involved in controversy. Its arrival on 26-Feb-1832 in Lavaca Bay was first reported from Barranco Colorado by Rafael Chowell, who informed Goliad that the vessel was importing goods for local merchant John Linn and two others, that he would travel to Goliad with the cargo, and the vessel intended to call next at Cópano [Chowell 26-Feb-1832]. It arrived there on 7-Mar-1832 with passengers, including Irish priest Father John Molly, two nuns (or monks/friars) and two youngsters in their care, all under a valid passport for settlements on the Nueces [Elosúa 27-Mar-1832]. José Bonifacio Galan (Commissioner at Goliad) became upset about the circumstances of the vessel's arrival, and he wrote to high-level authorities in Leona Victoria/Saltillo (greatly avoiding the local Texas chain-of-command) to complain, apparently since he had not been informed or invited to certify the arrival, but the authorities referred the matter back to Terán in Matamoros [Fernandez Rincón 24-Apr-1832]. Incredibly, though, even the Mexican federal Minister of Relations and the Governor of Coahuila and Texas got involved and issued orders about where to locate these few passengers of Cópano [Del Valle 7-May-1832, Músquiz 24-May-1832]. Thus, it appears a nasty "turf battle" ensued between Galan at Goliad, and the folks at Guadalupe Victoria and Barranco Colorado. It appears Terán later requested statements from Chowell and Arteaga, which were provided and in which both men indignantly refute the claims of and counter-charge Galan with stupidity and bad faith [Arteaga 12-Jul-1832 frames 554-556, Chowell 12-Jul-1832]. In all of this correspondence, the *Cañon* is often described as a *"Goleta Nacional"*, implying that its seizure back in 1830 probably resulted in it being taken into permanent ownership by the Mexican government.

It is not clear exactly what transpired next (that is, was it another movement along the coast or a subsequent voyage's return trip), and few details are given in available correspondence, but it was reported the *Cañon* *"fell to pieces in the Bar of Matagorda"* on or before 23-May-1832 [Cosío 30-May-1832], so its shipwreck is probably one of many yet to be found in the treacherous and shifting shoals at Paso Cavallo.

During this period of time, a letter from Terán to Arteaga indicates that approval had been granted for Arteaga's retirement [Terán 12-Apr-1832]; however, this would change due to ongoing events (such as those described above) and into the summer of 1832.

Arteaga chose to move his command from Barranco Colorado back to Guadalupe on 1-May-1832 *"to escape from the illnesses that I experienced last year"*, as well as scurvy and bad water. He also repeated his complaint that the cavalry troop which had been posted at Guadalupe from Goliad did not have any supplies, that he could not provide for them, and pleaded for better-supplied cavalry [Arteaga

15-May-1832]. In light of these events, Arteaga recognized that he had to postpone his retirement, and wrote to Elosúa about its delay, and also that he had received orders to withdraw from Lavaca [Arteaga 28-Jun-1832 frames 105-106]. The shortages extended to their salaries, and also gunpowder and bullets in the summer, and was eventually referred up the chain of command to Matamoros for aid [Arteaga 1-Jul-1832]. Arteaga was told he could request ammunition from the limited supplies at Goliad [Elosúa 30-Jun-1832].

Aniceto Arteaga learned of the Battle of Velasco and/or its prelude quickly, as there were travelers that passed that way, and they left statements that were preserved by the authorities at Lavaca and Guadalupe. One was an account by an express mail rider from Guadalupe to Brazoria, Romualdo Quintanilla, who met one Cayetano Garza traveling the other way (from Brazoria to Mier) while stopped at the Garcitas Creek crossing. Garza reported that 100 Americans were marching from Brazoria to Anahuac as of 17-Jun-1832, and that he (Garza) had met one Vicente Padilla (traveling from Velasco to Nacogdoches) at the San Bernard, and that Padilla had said the same number of Americans had disarmed 100 men at the mouth of the Brazos as of 16-Jun-1832 (although he'd not directly witnessed such) [Quintanilla 24-Jun-1832, Arteaga 24-Jun-1832]. Apparently, Quintanilla had ordered one of his accompanying soldiers (Miguel Lopez) to take Garza to the Lavaca Post, where both Lopez and Garza were further interviewed, at the specific request of Arteaga. The dates would indicate these men probably observed the uproar among the Brazoria colonists about the Anahuac Disturbances but BEFORE the actual Battle of Velasco on 26-27 Jun 1832.

Arteaga got confirmation that a true battle had occurred at Velasco on or before 2-Jul-1832, since he was provided with a Spanish translation of a letter sent by one of the Texians at the battle, James B. "Britt" Bailey, requesting reinforcements [Bailey 27-Jun-1832]. Apparently, the letter was translated by José M. J. Carbajal (probably at San Felipe de Austin) and then sent down to Arteaga, who copied it and sent the copy on to Elosúa at Béxar. Rámon Músquiz, the political chief then visiting at San Felipe, also wrote twice to the military commander at Lavaca with news of the Battle of Velasco, suggesting reinforcements [Músquiz 30-Jun & 2-Jul-1832].

The military responses of the Lavaca detachment to the Anahuac Disturbances and the Battle of Velasco are very difficult to reconstruct in modern times, due to the fact that many different commanders issued conflicting orders or dispatched forces often at a distance in terms of time and miles (some documents of which probably did not survive). These orders were also probably almost as confusing to the participants at the time. And, as discussed above, the Lavaca detachment had suffered many deprivations, difficulties and distractions. There were responses both before and after the Battle of Velasco, which evolved as new information became available. Here, an attempt to reconstruct the main activities will be made to the extent possible, based on the available surviving documents. <u>Before</u> the Battle of Velasco, apparently in response to the Anahuac Disturbances, Guerra issued orders from Matamoros (again subbing for Terán) to Elosúa for Tenoxtitlán, Lipantitlán and Goliad to reinforce Fort Velasco. A week later, Guerra rescinded those orders, apparently after hearing from Terán, who thought Stephen F. Austin should preferentially handle the matter [Guerra 20-Jun-1832]. Apparently, the original order also went directly to Lavaca, which caused Arteaga to coordinate with the commander at Goliad (Mariano Cosío), standing by with 1 officer and 24 men to join with forces from Goliad on their way to Velasco [Arteaga 28-Jun-1832 frames 107-108], which did depart in that direction. Belatedly, and perhaps against his better judgement (since he may have become aware of the Battle of Velasco), Cosío obeyed the superior order from Guerra and issued an express mail to recall troops from Velasco [Cosío 3-Jul-1832]. <u>After</u> the Battle of Velasco became known to him, Rámon Músquiz, the political chief of Texas then nearby at San Felipe, also wrote directly to Arteaga (twice), suggesting reinforcements for Velasco [Músquiz 30-Jun-1832]. Amid these conflicting orders and after recalling his group of 70 soldiers, Cosío convened a meeting with his officers on 5-Jul-1832, and they decided that a troop of 15 dragoons should be divided, ten going to Arteaga at Guadalupe to help manage the prisoners, and 5 staying at Goliad to maintain the mail routes. The mail to Ugartechea was held up since they feared it would be captured if delivered to Brazoria/Velasco [Cosío 6-Jul-1832]. Although Ugartechea wrote to Arteaga on 8-Jul-1832 (from San Felipe) that the Velasco situation was all settled and that they were already headed back to Mexico, which Arteaga did not receive until the 10th [Ugartechea 8-Jul-1832], Arteaga was still planning to send troops to aid Ugartechea on the 9th based on Músquiz's request [Arteaga 9-Jul-1832]. Despite the confusion, there was really no point in reinforcing Velasco at this time (as Ugartechea had abandoned it and was already in San Felipe), so it is surmised all troops eventually returned to their home bases.

The last extant documents sent by Arteaga from Lavaca appear to be a bundle of six letters on 12-Jul-1832 [Arteaga 12-Jul-1832], so it is surmised that he delayed his retirement until the withdrawal of his command from Texas was complete. One of these dispatches was a cover letter for a military report, although the report itself appears to be absent from the Béxar Archives. Another is a lengthy report on accusations surrounding the schooner **Cañon** and other ships that had called at Lavaca. Chowell also wrote a similar letter, indicating that a significant feud had developed with the Commisssioner of Goliad (José Bonifacio Galan) [Chowell 12-Jul-1832], as discussed more fully above. Arteaga did receive some ammunition, brought by the 10 dragoons which came from Goliad, in the form of 1500 cartridges [Cosío 13-Jul-1832]. Rafael Chowell wrote a cover letter for a budget report in July to Erasmo Seguin, still seeking financial assistance or loan for Lavaca's accounts, but the attached report is also missing [Chowell 15-Jul-1832].

During this two-year period, there are many mentions of individual desertions or deaths in the Lavaca detachment, too numerous to mention here, found in the Béxar Archives in specific dispatches or in the side notes of the monthly military reports. Indeed, in the graphs shown in Figures 28 and 29, a gradual decline can be seen in the complement of each unit, reflecting this attrition. It would seem likely that deaths among the soldiers and prisoners resulted in burials in a nearby small cemetery, although no direct mention of such has been found. The population of the 12th Battalion and the prisoners ended up being about half of the starting amount. Illness, desertions, lack of funds for salaries and provisions, the isolated location and long-distance communication from many commanders must have made it almost impossible to accomplish the goals that Terán had in mind. And, then Terán committed suicide on 3-Jul-1832 behind a church in Padilla, Tamaulipas near his new headquarters, already ill and overworked, despondent over Mexican politics (since he had sided with the unsuccessful centralist regime that had just fallen to Santa Anna) and his belief that Texas was lost.

The details surrounding the actual departure of Arteaga from the Lavaca area are a bit murky but seem to involve a bit of intrigue. It seems that Arteaga unexpectedly declared support for Santa Anna on 4-Aug-1832 and fled to "*Garcitas*" with the majority of his command (6 officers and 70 soldiers) [Moret 6-

Aug-1832], where he embarked for Tampico through the mouth of Matagorda Bay [Hernandez 1832]. The author of the former letter (Juan Moret) was a junior officer known to have been in Ugartechea's command at the Battle of Velasco, and is presumed to have been traveling overland back to Matamoros (being then in Guadalupe) where he took some responsibility to command the abandoned post, although he intended to continue travel on to Matamoros with a group of troops remaining loyal to the Bustamante government. He further reports that his *"Jefe primer Ayudante"* also went with Arteaga and that Lt. Miguel Nieto had disappeared. In the Mexican custom, Arteaga even penned a "pronunciamiento" sent to Elosúa [Elosúa 16-Aug-1832], and that, pursuant to orders from José Antonio Mexía, he embarked with his troops. The original document itself seems to have been forwarded with this message to Terán's replacement (thought to have been Ignacio de Mora), and appears absent from the Béxar Archives. Rafael Chowell apparently did not join Arteaga, and instead traveled to Goliad where he confirmed Arteaga's pronunciamiento, and sought instructions from Elosúa and the Commissioner at Matamoros [Chowell 10-Aug-1832].

It is known, of course, that Mexía (commander of the Tampico garrison) had sided with the Santa Anna party, and had left Matamoros with 6 ships (3 schooners, 1 brig and 2 smaller vessels [Guerra 27-Jun-1832 frames 75-76]) and 400 men in response to the Battle of Velasco and the Anahuac Disturbances, and then arrived off the mouth of the Brazos River on 16-Jul-1832, staying about a week in the Velasco/Brazoria area [Cotten 1832, Turner 1903 pp. 12-13]. In attempts to recruit the Anahuac garrison to his cause, Mexía and his ships then traveled by sea to Galveston Bay on or about 24-Jul-1832 where they met members of this garrison aboard two or three ships as they were leaving Galveston Bay, under the command of Felix Subarán declaring for Santa Anna [Mexía 18-Jul-1832, Turner 1903 p. 14-15], after which they all returned by sea to Matamoros and then to Tampico. It is also known that military commanders along the Texas coast had been warned of Mexía's *"dissident flotilla"* in advance [Guerra 27-Jun-1832]. So, given the coincidental timing, it is possible that covert dispatches were exchanged between Arteaga and Mexía in late Jul-1832, and arrangements were made for one or more ships to secretly pick up Arteaga at Lavaca Bay on Mexía's return down the coast. Some details might be found by recalling the words of John Linn that ". ... *After the surrender of Velasco and the intrigues of Santa Anna had been developed, and after the death of General Teran, who committed suicide by falling upon his sword, Commissioner Choval resigned, and Captain Artiaga informed General Mexia that he wished to be relieved of the responsibilities of his position, as he did not favor the movements of Santa Anna. An order arrived directing the removal of the whole army, together with the workmen at the brick-kiln, some thirty or forty in number. .... Lieutenant-Colonel Villasana arrived in a schooner in the bay to transport the troops to Matamoros …. Captain Artiaga called on me and stated that he was ordered to abandon the proposed fort; that he needed supplies, and that Villasana would draw on the Matamoros custom-house for the same.*" Linn's words suggest that Arteaga was opposed to the Santa Anna party, but perhaps he had a change of heart in the summer of 1832. Arteaga was remembered in Guadalupe Victoria, as one of the old downtown streets was named after him, now known as Forrest Street. Rafael Chowell also had a street named for him (as Choval Street), now Constitution Street. These two parallel east-west streets bracketed a city block called "Plaza de Constitution" originally, then considered the center of Guadalupe Victoria, now known as "DeLeon Plaza" [Shook 2007 p. 473].

It will not be considered unusual that both the Tejano and Anglo-American colonists in and around Guadalupe Victoria also sided with the Santa Anna party (then supposedly supporting the Mexican

Constitution of 1824), and that specific meetings were held there in support of this "liberal" or "federalist" party (one on July 16th, which elected a committee of vigilance) [Turner 1903, pp. 17-18]. What is not known, is what influence these discussions may have had upon Arteaga and his command, then stationed in the same small town. Arteaga and his men had suffered illness, as well as lack of salary, funds, provisions, ammo and other supplies, the **Cañon** controversy, the death of Terán, and abandonment of Fort Velasco and Anahuac – all of which may have caused them to now side with the Santa Anna party. However, it does appear that Arteaga secretly fled with the majority of his command at some point in early August of 1832, perhaps posting his "pronunciamiento" as he left (received by Elosúa on the 16th), as he was still present on the 4th when Moret learned of his declaration. And, as we have seen, Arteaga was not alone in joining with the Santa Anna party (then achieving more success in its revolution in Mexico), and helps to explain the rapid abandonment of forts at Velasco, Anahuac, Nacogdoches and Lavaca that summer. In a larger sense, the resulting lack of military control of Texas, followed by its harsh reapplication in 1835, were important antecedents to the Texas Revolution of 1835-1836.

Two years after it was abandoned, the location of Barranco Colorado was shown on a chart thought to have been drawn by Jean Louis Berlandier in 1834, shown in Figure 31 below [Berlandier 1834]. He was a naturalist who had also accompanied Terán's boundary expedition to Texas in 1828, and passed through this area in 1829 on an excursion from Béxar by land to La Bahía, Cópano, and then by sea to New Orleans, returning the same way. He returned to La Bahía (Goliad) in 1834, probably drawing this map of his route while traveling in that vicinity. The chart shows "La Vaca" (red oval) on "Arroyo de La Vaca", which the caption says was named "Barranco Colorado" (green oval), and that a largely east-west road existed between "Guadalupe Victoria" and "La Vaca".

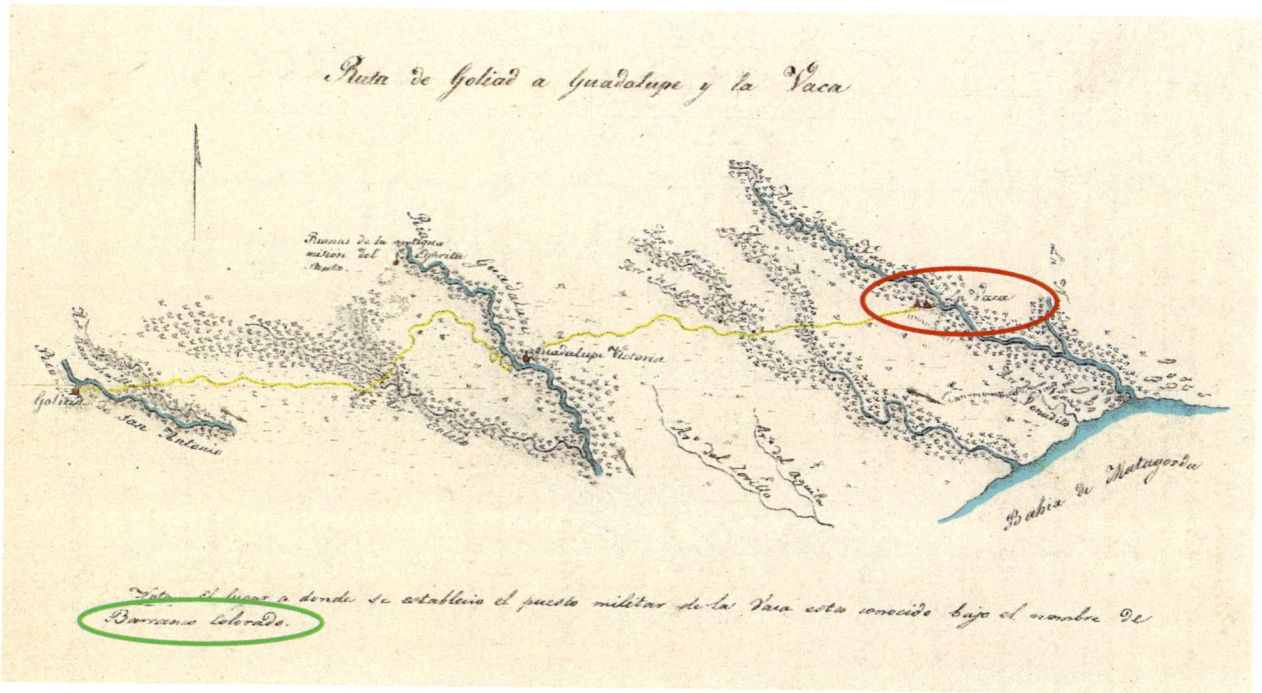

Figure 31: "Ruta de Goliad a Guadalupe (Victoria) y la Vaca", ascribed to Jean Louis Berlandier, circa 1834 [Berlandier 1834] – color & clarity enhanced version

Caption translation: The place where the military post of La Vaca was established is under the name of Barranco Colorado

Dr. Robert W. Shook concluded that this map was drawn in 1829, when Terán (and presumably Berlandier) traveled on their return trip through Guadalupe Victoria and La Bahía to Matamoros [Shook 2007 pp. 342-352]. However, we know from Berlandier's journal that he came instead from Béxar to La Bahía, so he could not have traveled through Barranco Colorado, especially since it did not yet exist. It seems more likely that Terán traveled down the San Antonio Road to the Trinity crossing, and then on the more-southerly Coushatta Trace to San Felipe, before continuing on to Guadalupe Victoria and La Bahia and eventually Matamoros. As indicated in the book "Texas by Terán" on pages 148 and 154, those authors conclude that the Berlandier map (Figure 31) was drawn by him during or after a later visit in 1834.

The 1830-1836 versions of the Austin/Terán/Tanner map (as seen in the Frontispiece, Figures 1 and 3) show a road directly from Victoria to San Felipe, and is drawn taking a more-northerly direction out of Victoria than shown in the Berlandier map of Figure 31. Dr. Shook labels this road as the San Felipe-Atascosito Road – it passed north out of Guadalupe Victoria, then bearing northeast to San Felipe, and from there through Harrisburg to the lower Trinity River. Another map of Texas came out in 1839 (by Richard S. Hunt and Jesse F. Randel) which has a more accurate rendering of Labaca (Lavaca) and Matagorda Bays, and especially so for Espiritu Santo (modern San Antonio Bay), Aranzazua Bay (modern Aransas Bay) and Copano Bay (please see Figure 32). The town of Texana is incorrectly shown on the left (east) bank of the Lavaca River, when it was really about 3 miles east on the right (west) bank of the Navidad River; however, it marks the approximate location of Barranco Colorado (which was on the opposite or right bank). As in Figure 31, a largely east-west road is shown from Victoria, which has a north fork bearing northeast to San Felipe, and a south fork or detour continuing east through the Barranco Colorado area and into Texana. This is very similar to the road segments shown in the 1858 Pressler map (Figure 18) and 1838 Linn map (Figure 19). Thus, the appearance of Barranco Colorado and then Texana caused a southern shift of the Atascosito Road east of Guadalupe Victoria. The Hunt-Randel map is also notable for showing the locations of many early short-lived towns such as Cópano, Aransas (modern Fulton area), Lamar, Calhoun, Linnville, Dimmits (Dimmit's Landing), Cox's Point and others.

Figure 32: Detail from 1839 Hunt-Randel map of "Texas"

After 1832, Barranco Colorado was abandoned, while the towns of Texana and Red Bluff grew up nearby. If the structures of the Mexican fort were made of wood, as suggested in the ***Texas Gazette*** newspaper (Figure 20), any evidence may have decayed before the area became settled, and knowledge of the site was "lost to history". Primary documents make little or no mention of brick-making at Barranco Colorado, let alone being shipped elsewhere.

There is a database or registry, maintained by the Texas Archeological Research Laboratory (TARL) in Austin, which formally records archaeological survey results and observations of potential archaeological sites throughout the state, typically reserved for use by professional archaeologists. One such reported site in Jackson County is known as 41JK29 (41 = Texas, JK = Jackson County, 29 = the 29th reported site in that county). Since brick was reported present, and the location is essentially the same as shown above for Barranco Colorado, it is concluded that the reported site must be associated with the 1830-1832 Mexican fort, despite the fact that the original reporter of the site (in 1967) failed to recognize or report it as such. The meager data reported for this site is shown below in Figure 33.

Site Description	Open area on terrace above the Lavaca Rv.-Plowing has turned up lots of bricks-Neighbor boys throw in river.
Area of Occupation	Bricks-40 by 40 ca. (Frank: this doesn't make sense)
Present Condition	Has been plowed for years
Character and Depth of Fill	Bricks found about 1 ft subsurface (now disturbed and on surface)
Previous Designations for Site	Old Brick Factory, Blair community (I.T. Taylor) Frank: I.T. Taylor's Book, *the Cavalcade of Jackson County*, is incorrect. The Blair community was quite distant from this site.)
Date	7/25/1967

Figure 33: Information from TARL database for 41JK29

A nearby landmark observed on older USGS maps is labeled as "Brick Factory Springs", which are described by Gunnar Brune: *"The Brick Factory Springs were on the west bank of the Lavaca River ten kilometers south of Edna, on James Reid's farm at latitude 28° 52' and longitude 96° 39'. The Mexican army was stationed here in 1831 and established a brick factory, using the red clay from the 16-meter high bluff. Below the clay beds and close to the river are silt and sand beds from which the small springs once flowed. The recharge area is one kilometer to the west, where these beds crop out at the surface. The springs have been dry for many years. Nearby some bones of a mammoth, such as were hunted by the Paleo-Indian people, were found."* [Brune 1981 pp. 253-254]. These springs seem to fit the description of fresh healthful water asked for in Terán's initial directives, while the face of the bluff offered clay-bearing strata for use in brick-making.

Based on all of the information above, the location of Barranco Colorado is surmised to be about 6 miles south of Edna, Texas, and is illustrated on a 1952 USGS map below in Figure 34, in concert with other historical markers found in the immediate area. A bluff approaching 50' elevation stands directly above the river at this point, with prairie covering its top, and an adjacent bottom show a wooded terrace at about 20' elevation. An intermittent creek draining from the north cuts through from higher elevation and then runs at the base of the bluff before emptying into the river, and the springs must emanate from the base of the bluff into this creek. The bluff so marked seems to be the most likely location, although the fort itself may have been located some distance away from the river on the higher flat prairie to maintain an open defensive perimeter and to avoid floods. The brick operation may have been close to the river, where the wooded bottoms provided fuel for the kiln/cooking/heat and timber for structures.

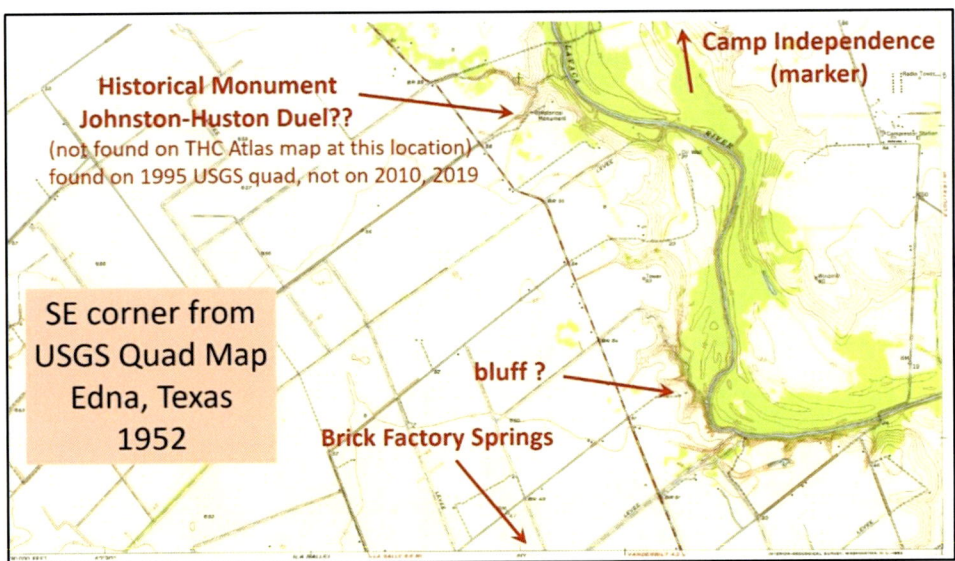

Figure 34: Likely Location of Barranco Colorado using 1952 USGS Map as Basis

A survey of the Lavaca and Navidad watershed was conducted in 1938 by the U.S. Army Corps of Engineers, which is documented with charts found in the National Archives. Sheet 4/Plate 21 (https://catalog.archives.gov/id/148373410) involves this area of the Lavaca River, and shows additional information with some elevation transects and topography. A portion of Sheet 4 is shown below as Figure 35. The north direction is atypical, but a red arrow shows the same bluff and creek at its base.

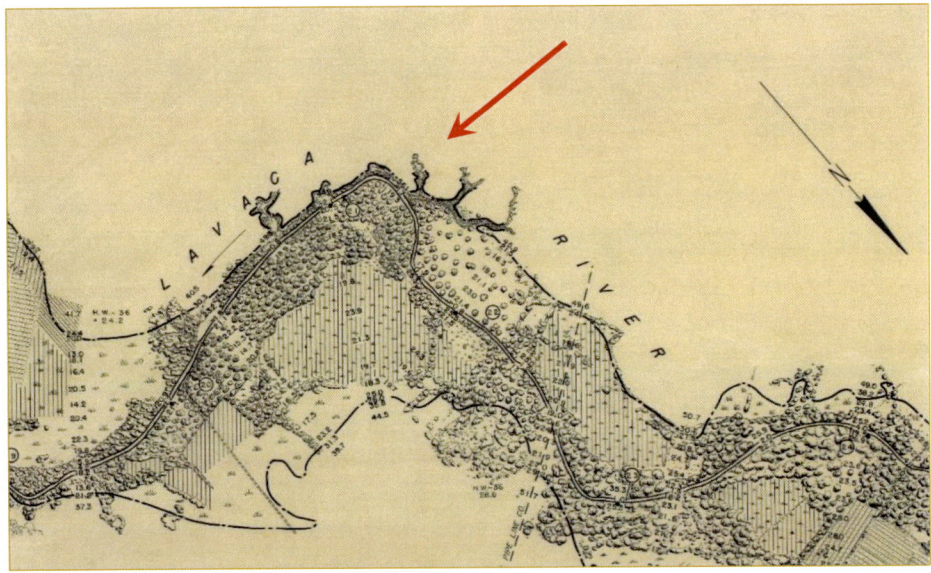

Figure 35: Portion of Plate 4 from 1938 USACE survey of Lavaca and Navidad River watershed

It seems odd that Barranco Colorado was lost to history since it was occupied for perhaps one and a half years where bricks were made, yet no substantial structural remains of a fort were left (as at Anahuac), especially when compared with the 1832 Fort Velasco which was occupied for only about two months.

CONCLUSIONS

1. Barranco Colorado was the name for an isolated post on the lower Lavaca River, as part of the implementation of the Mexican Law of 6-Apr-1830 and under the overall direction of Manuel de Mier y Terán. Its military commander was Aniceto Arteaga, and its commissioner was Rafael Chowell. The post was not established immediately upon arrival of Arteaga and his infantry troops in the summer of 1830, but was occupied in some form or fashion from Oct-1830 to May-1832, after which it was abandoned. Barracks and a brick kiln were apparently built at the site, although it does not seem that any significant brick structures, especially a stronghold or fort, was completed there.
2. Establishment of this post appears to have replaced the need for a small detachment or camp placed near Paso Cavallo known as "Matagorda" or "Port of Matagorda", which had been manned by troops out of La Bahía in the years prior, say 1826 to 1830.
3. No evidence was uncovered in primary documents to suggest that the brick-making operation was substantial, or that bricks were actually shipped off-site.
4. The existing community of Guadalupe Victoria, established in the 1820's under the Mexican empresario Martin de Leon as the headquarters of his Guadalupe colony, was used as an initial base of operations by Arteaga. Once Arteaga and his infantry moved to Barranco Colorado in Jan-1831, a large cavalry unit under José Manuel Barberena was garrisoned at Guadalupe Victoria until Nov-1831. Arteaga and his detachment returned there in May-1832 until they departed Texas in Aug-1832.
5. The combined command at Guadalupe and Barranco Colorado appears to have been referred to as "*Comandancia Militar del Establecimiento de la Vaca*", translated as "Military Command of the Establishment of Lavaca". Its headcount exceeded 200 troops for a period of some months in the latter half of 1831, but was less at other times. This detachment stood astride the Guadalupe and Lavaca Rivers near their mouths, in an attempt to control access to that part of the Texas coast.
6. The detachment suffered from lack of provisions, lacks of funds and pay, disease, desertions and deaths – making it very difficult to succeed in the purposes imagined by Terán. Due to the number of deaths, it is also likely that a small cemetery was located in the area of Barranco Colorado.
7. Lavaca was one of the larger detachments among the six new forts that Terán commissioned in 1830-1832. Yet, despite this, its presence is an under-told aspect of the era's history. Together with the failure of the other Terán forts, it was an antecedent of the Texas Revolution of 1835-1836.
8. And, finally, Barranco Colorado capped a long, difficult period for "Bahía de San Bernardo" during the colonial period, many elements of which have been lost to history but bear remembering.

RECOMMENDATIONS

1. A formal historical marker does not exist for Barranco Colorado or Comandancia Militar del Establecimiento de la Vaca. For the former, a location on a nearby public road (for example, along FM-234 near its intersection with Jackson County Road 320) might be appropriate. Such an effort might be considered by the local county historical commissions.
2. If permission can be obtained from the private landowner of 41JK29, modern techniques for geophysical surveying (such as ground-penetrating radar or magnetometry) might be useful to identify the extent or exact locations of any remaining portions of Barranco Colorado.

People who take no pride in the noble achievements of remote ancestors will never achieve anything worthy to be remembered with pride by remote descendants.
   - *Lord Thomas Babington Macaulay*

REFERENCES

Collections (in alphabetical order by abbreviations - used afterwards in references)

AP - Barker, Eugene C. (editor), The Austin Papers, Vol. I (2 parts, Annual Report of the American Historical Association For The Year 1919, United States Government Printing Office, Washington DC, 1924), Part 1 (1789-1824) and Part 2 (1825-1827); *confusingly, the spine and title page of the original volumes label it as Vol. II, Parts 1 or 2, although there does not seem to have been a separate published Volume I;*
Vol. II, 1828-1834, (Annual Report of the American Historical Association for the Year 1922, United States Government Printing Office, Washington DC, 1928); Vol. III (Oct, 1834-Jan, 1837), The University of Texas Press, Austin, Texas, 1926
AGMC - Archivo General de Mexico Collection, Dolph Briscoe Center for American History, The University of Texas at Austin, Austin, Texas
BA - Béxar Archives, Dolph Briscoe Center for American History, The University of Texas at Austin, Austin, Texas
GFP - George Fisher Papers, 1830-1948, Dolph Briscoe Center for American History, The University of Texas at Austin, Austin, Texas
JLBP - Jean Louis Berlandier Papers, WA MSS S-300, Beinecke Rare Book And Manuscript Library, Yale University, New Haven, Connecticut
M&SFAP - Moses and Stephen F. Austin Papers, Dolph Briscoe Center for American History, The University of Texas, Austin, Texas
PCRCT - *Papers Concerning Robertson's Colony in Texas* (edited by Malcolm D. McLean), University of Texas at Arlington Press
PTTH – The Portal to Texas History, The University of North Texas Libraries
RBBRC – Robert Bruce Blake Research Collection, compiled in the Eugene C. Barker Texas History Center Archives Collection, 1958-1959 (*full set available at Clayton Library, Houston, Texas*)
SMWC - Samuel May Williams Collection, Galveston & Texas History Center, Rosenberg Library, Galveston Texas
SWHQ - *The Southwestern Historical Quarterly* (journal of The Texas State Historical Association), 1912-present; including previous version *The Quarterly of the Texas State Historical Association*, 1897-1912
TGLO – Texas General Land Office, Spanish Archives (Austin, Texas)

Ahumada, Mateo,
1-Feb-1826	Letter to Stephen F. Austin, **BA**, Microfilm Roll 89, frames 76-78; see also Ahumada to Rojo & Ahumada to Austin, 10-Apr-1826, Roll 91, frames 470-473 & 474-479

Almonte, Juan (author),
Jan-1925 & 2003	Carlos E. Castañeda (translator), "Statistical Report on Texas", **SWHQ**, Vol. XXVIII (28), No. 3, pp. 177-222; also see *Almonte's Texas* (Jack Jackson, editor & John Wheat, translator), Texas State Historical Association, p. 238-240

Arévalo, Mariano (publisher),
1829	*Colección de Ordenes y Decretos de la Soberana Junta Provisional Gubernativa, y Sobernas Congresos Generales de la Nación Mexicana*, (Mexico City), Vol. IV (4), p.6

Arteaga, Aniceto,
1-Jul-1830	Letter to Antonio Elosúa, **BA**, Microfilm Roll 132, frames 115-116
6-Jul-1830	Letter to Stephen F. Austin, **M&SFAP**, Series IV, 1830; typescript copy at **PTTH**: https://texashistory.unt.edu/ark:/67531/metapth217433/, accessed 17-Jan-2021
1-Aug-1830	Monthly Military Report, **BA**, Microfilm Roll 133, frame 24
19-Aug-1830	Letter to Antonio Elosúa, **BA**, Microfilm Roll 133, frames 678-680
25-Aug-1830	Letter to Antonio Elosúa, **BA**, Microfilm Roll 133, frames 788-797 (especially 797); see also Elosúa to Terán (31-Aug-1830) frames 934-937 (especially 936), Terán to Elosúa (22-Sep-1830) Roll 134, frames 567-578 (especially 574), Terán to Seguin (22-Sep-1830) frames 586-592 (especially 590-591)
1-Oct-1830	Monthly Military Report, 1-Oct-1830, **BA**, Microfilm Roll 134, frame 934
12-Oct-1830	Letter to Antonio Elosúa, **BA**, Microfilm Roll 135, frames 350-353
12-Jan-1831	Letter to Antonio Elosúa, **BA**, Microfilm Roll 137, frames 782-783; see also Elosúa to Arteaga (18-Jan-1831), Roll 137, frame 7
22-Jun-1831	Letter to Antonio Elosúa, and Elosúa's reply, **BA**, Microfilm Roll 142, frames 158-161 (especially 159-161)
28-Jun-1831	Letter to Antonio Elosúa, **BA**, Microfilm Roll 141, frames 762-763; see also Arteaga to Elosúa (29-Jun-1831) Roll 142, frames 366-367, Arteaga to Elosúa (27-Jul-1831) Roll 143, frames 157-160, Elosúa to Cosío & Arteaga (2-Aug-1831) Roll 143, frames 455-457, Arteaga to Elosúa (7-Aug-1831) Roll 152, frames 403-404, Arteaga to Elosúa (27-Aug-1831) Roll 143, frames 997-999
6-Jul-1831	Letter to Antonio Elosúa, **BA**, Microfilm Roll 142, frames 625-626; see also Arteaga to Elosúa (9-Jul-1831), Roll 142, frame 713, Chowell's list of goods (27-Jul-1831), Roll 143, frames 161-162, Arteaga to Elosúa (28-Jun-1831), Roll 152, frames 2-3, Arteaga to Elosúa (7-Aug-1831), Roll 143, frame 564, Cuellar's list of goods (16-Aug-1831), Roll 143, frame 774, Arteaga to Elosúa (20-Aug-1831), Roll 143, frames 901-902 and Elosúa to Arteaga (30-Aug-1831), Roll 143, frame 903
11-Dec-1831	Letter to Antonio Elosúa, **BA**, Microfilm Roll 154, frames 254-255
11-Apr-1832	Letter to Antonio Elosúa, **BA**, Microfilm Roll 149, frames 333-334
15-May-1832	Letter to Antonio Elosúa, **BA**, Microfilm Roll 150, frames 26-28
24-Jun-1832	Letter to Military Commander of Lavaca (and reply), English translation, RBBRC, Vol. XII, pp. 353-358
28-Jun-1832	Letter to Antonio Elosúa, **BA**, Microfilm Roll 151, frames 100-108

1-Jul-1832 Letter to Antonio Elosúa, **BA**, Microfilm Roll 151, frames 226-237; also see Elosúa to Arteaga (5-Jul-1832) Roll 151, frame 377
9-Jul-1832 Letter to Antonio Elosúa, **BA**, Microfilm Roll 151, frames 475-476
12-Jul-1832 Letter to Antonio Elosúa, **BA**, Microfilm Roll 151, frames 548-556

Austin, Stephen F.,
27-May-1823 Letter to Felipe de la Garza, transcript in **AP**, Vol. I, pp. 651-653; *at Digital Austin papers:* http://digitalaustinpapers.org/document?id=APB0597
1-Oct-1824 Letters to Supreme Executive Power of the Mexican Republic, 1-Oct-1824 and 6-Nov-1824, ... State Congress, 6-Nov-1824, ... Governor of Coahuila and Texas, 4-Feb-1825, ... Governor Rafael Gonzales, 4-Apr-1825), transcripts in **AP**, Vol. I, pp. 912-913, 935-936, 936, 1034-1035, 1065-1067; *latter two are in English and found at Digital Austin Papers*: http://digitalaustinpapers.org/document?id=APB1035 and http://digitalaustinpapers.org/document?id=APB1071
18-Mar-1826a Letters to José Antonio Saucedo & Erasmo Seguin, transcripts in **AP**, Vol. I, pp. 1281-1283 and 1288-1290 (copy of letter to Saucedo is in **TGLO**, Document# 730, Box 126/3, p. 19); see also letters of Austin to Emily Perry, 28-Jan-1826 (**AP** I:1260-1262) and Austin to Saucedo, 27-Mar-1826 (**AP** I:1299)
18-Mar-1826b Letters to Rafael Manchola & Mateo Ahumada; also see Ahumada to Austin, 24-Jul-1826, transcripts in **AP**, Vol. I, pp. 1285, 1285-1288 and 1155-1156
24-Dec-1829 Letter to José Antonio Navarro, transcript in AP, Vol. II, pp. 302-303; at Digital Austin Papers: http://digitalaustinpapers.org/document?id=APB1816
13-Mar-1830 *Notice and editorial published in Texas Gazette issues of 13- and 27-Mar-1830,* transcript of latter in **AP**, Vol. II, p. 351; at Digital Austin Papers: http://digitalaustinpapers.org/document?id=APB1884, accessed 19-Jan-2021
29-Mar-1830 Letter to Rámon Músquiz, transcript in **AP**, Vol. II, pp. 354-355
13-Jul-1830 Letter to Aniceto Arteaga, transcript in **AP**, Vol. II, p. 451

Bache, Alexander D., S. A. Gibert, M. Seaton, J. C. Febiger,
1857 *Preliminary Chart Of Entrance To Matagorda Bay, Texas*, From a Trigonometrical Survey under the direction of A. D. Bache Superintendent of the Survey Of The Coast Of The United States

Bagby, Mindora,
22-Oct-1924 "The Local History of Jackson County", University of Texas Bulletin, No. 2440, The Texas History Teachers' Bulletin, Vol. XII, No. 1, pp. 74-78

Bailey, James B.,
27-Jun-1832 Letter to David Shelby, **BA**, enclosure in Microfilm Roll 151, frames 341-344; English translation in **RBBRC**, Supplement Volume IX, pp. 266-267; the Bailey letter at **BA** is a copy by Aniceto Arteaga on 2-Jul-1832 of a Spanish translation by José M. J. Carbajal on 29-Jul-1832 of the original English letter, and then enclosed in a letter from Aniceto Arteaga to Col. Antonio Elosúa on 4-Jul-1832. The Spanish-language Carbajal translation is found in the Nacogdoches Archives, Vol. 62, pp. 194-195 (and its translation back into English is also in **RBBRC**, Vol. XII, pp. 369-370

Barker, Eugene C.,
1926	*The Life of Stephen F. Austin, Founder of Texas, 1793-1836, A Chapter in the Westward Movement of the Anglo-American People* (University of Texas Press, Austin and London)

Barberena, José Manuel,
1-Aug-1830	Monthly Military Report, **BA**, Microfilm Roll 133, frames 10-13
12-Jun-1831	Letter to Antonio Elosúa & reply of 16-Jun-1831, **BA**, Microfilm Roll 141, frames 889-890
9-Nov-1831	Letter to Antonio Elosúa, **BA**, Microfilm Roll 145, frames 883-885

Berlandier, Jean Louis,
1829	Map entitled *"Bahia de San Bernardo ou Bahia de Matagorda"*, **JLBP**, Box 8; available on-line as 5th image at https://collections.library.yale.edu/catalog/2028735 ; accessed 21-Nov-2022
1834	Map entitled *"Ruta de Goliad a Guadalupe (Victoria) y la Vaca"*, circa 1834, **JLBP**, Box 8, Volume II, p. 4r; available on-line at:
https://collections.library.yale.edu/catalog/32497677?child_oid=32497774; last accessed 8-Feb-2023
1980	*Journey to Mexico During the Years 1826 to 1834*, (two volumes, translated by Sheila Ohlendorf, Josette M. Bigelow and Mary M. Standifer, introduction by C. H. Muller), Texas State Historical Association

Brune, Gunnar,
1981	*Springs of Texas*, Vol. 1, 1st Edition, 1981, Fort Worth, Texas

Bryan, James P., Walter K. Hanak,
1961	*Texas in Maps* (1961, University of Texas, Austin, Texas)

Bugbee, Lester G.,
Oct-1899	"What Became Of the Lively?", **SWHQ**, Vol. III (3), No.2, p. 141-148

Castillo, José María,
3-Jul-1830	Commissary Review Reports for 12th & 11th Permanent Battalions at Guadalupe, **BA**, Microfilm Roll 132, frames 229-231 & 232-234
3-Aug-1830a	Commissary Review Report for 11th Permanent Battalion at Guadalupe, **BA**, Microfilm Roll 133, frames 141-142
3-Aug-1830b	Commissary Review Report for 12th Permanent Battalion at Guadalupe, **BA**, Microfilm Roll 133, frames 143-145
3-Jan-1831	Commissioner review report for 12th Permanent Battalion at Barranco Colorado, certified by Arteaga and Chowell, **BA**, Microfilm Roll 137, frames 591-594

Chowell, Rafael,
10-Jul-1830	Letter to Stephen F. Austin, 10-Jul-1830, **M&SFAP**, Series IV; typescript copy available at **PTTH**: https://texashistory.unt.edu/ark:/67531/metapth217440/m1/1/, last accessed 17-Jan-2021
3-Aug-1830	Certificates of Attendance, **BA**, Microfilm Roll 133, frames 146-147

19-Aug-1830	Letter to Erasmo Seguin, **BA**, Microfilm Roll 133, frames 672-675
12-Oct-1830	Letter to Erasmo Seguin, **BA**, Microfilm Roll 135, frame 347
8-Nov-1830	Letters to Antonio Elosúa, 8- and 17-Nov-1830, English translation in **PCRCT**, Vol. 5, 1978, p. 164 & 219; Spanish originals found in **BA**, Microfilm Roll 136, frames 123 & 312
26-Feb-1832	Letter to José Bonifacio Galan, **BA**, Microfilm Roll 148, frame 273; also see Cisneros to Galan (9-Mar-1832) frames 543-544
12-Jul-1832	Letter to Antonio Elosúa, **BA**, Microfilm Roll 151, frames 557-559
15-Jul-1832	Letter to Erasmo Seguin, **BA**, Microfilm Roll 151, frames 637-638
10-Aug-1832	Letter to Antonio Elosúa, **BA**, Microfilm Roll 152, frames 465-466

Clay, Comer

Apr-1949	"The Colorado River Raft", **SWHQ**, Vol. LII (52), No. 4, pp. 410-426

Cosío, Mariano,

26-Feb-1830	Letter to Antonio Elosúa, **BA**, Microfilm Roll 128, frames 702-710
23-Apr-1830	Letters to Antonio Elosúa, **BA**, Microfilm Roll 129, frames 928-932; see also Elosúa to Terán (27-Apr-1830) Roll 130, frames 60-64 (especially 60), Elosúa to Cosío (27-Apr-1830) frames 65-72 (especially 66), Terán to Elosúa (26-May-1830) frames 721-738 (especially 737-738), Cosío to Elosúa (2-Jul-1830), Roll 132, frames 136-155 (especially 153-154), Elosúa to Terán (5-Jul-1830) frames 301-306 (especially 303) and Elosúa to Terán (6-Jul-1830) frames 363-375 (especially 363)
6-Jun-1830	Letter to Antonio Elosúa, **BA**, Microfilm Roll 131, frames 371-376
12-Jun-1830	Letter to Antonio Elosúa, **BA**, Microfilm Roll 131, frames 596-599; also see Cosío to Elosúa (17-Jun-1830) frames 697-708 (especially 697-701), Aldrete to Músquiz (17-Jun-1830) frames 709-722 (especially 721b-722b), Cosío to Elosúa (18-Jun-1830) frames 743-756 (especially 751-754), Elosúa to Cosío (21-Jun-1830) frames 823-836 (especially 833-834) , Elosúa to Terán (6-Jul-1830) frames 363-375 (especially 367), Terán to Elosúa (16-Aug-1830) frames 565-574 (especially 567-568)
17-Jun-1830	Letter to Antonio Elosúa, **BA**, Microfilm Roll 131, frames 697-708 (especially 702-706); see also Cosío to Elosúa (12-Jun-1830) frames 596-599, Cosío to Elosúa (29-Jun-1830) Roll 132 frames 37-41, Cosío to Elosúa (2-Jul-1830) frames 136-155 (especially 140-142 & 151-152), Elosúa to Terán (5-Jul-1830) frames 301-306 (especially 304-305), Aldrete to Músquiz (12-Aug-1830) Roll 133 frames 426-443 (especially 434-435b)
2-Jul-1830	Letter to Antonio Elosúa, **BA**, Microfilm Roll 132, frames 136-155 (especially 145-146)
30-Jul-1830	Letter to Antonio Elosúa, **BA**, Microfilm Roll 132, frames 938-960
13-Aug-1830	Letter to Antonio Elosúa, **BA**, Microfilm Roll 133, frames 475-490
30-May-1832	Letter to Antonio Elosúa, **BA**, Microfilm Roll 150, frames 307-313 (especially 307-308); see also Elosúa to Terán (5-Jun-1832) frames 573-577 (especially 574-575) & Elosúa to Cosío (5-Jun-1832) frames 578-600 (especially 586-587)
3-Jul-1832	Letter to Antonio Elosúa, **BA**, Microfilm Roll 151, frames 287-290
6-Jul-1832	Letter to Antonio Elosúa, **BA**, Microfilm Roll 151, frames 381-390
13-Jul-1832	Letter to Antonio Elosúa, **BA**, Microfilm Roll 91, frames 587-588; see also De La Garza to Elosúa (13-Jul-1832) Roll 151 frames 598-599

Cotten, Godwin B. M. (editor and publisher),

23-Jul-1832	*Texas Gazette And Brazoria Commercial Advertiser newspaper*, EXTRA issue, copy of original and also transcription courtesy of BCHM

De Evia, José,
1785   Map, Plano de la Bahia de San Bernardo, Portulano de la Admiral Setentrional, Mapoteca Orozco y Berra (Mexico City), digitized color image available on-line at https://mapoteca.siap.gob.mx/index.php/coyb-int-m50-v3-0101/ , accessed 21-Oct-2022

De La Garza, José María,
7-May-1830   Dispatch about *Oposición*, **BA**, Microfilm Roll 130, frames 595-596; see also Galan to Elosúa (7-May-1830) frames 288-292, Elosúa to Cosío (10-May-1830) frames 388-391, Elosúa to Galan (11-May-1830) frame 421, Cosío to Elosúa (21-May-1830) frames 591-594, Elosúa to Terán (29-May-1830) frame 798, Elosúa to Cosío (29-May-1830) frame 799, Cosío to Elosúa (4-Jun-1830) Roll 131 frames 304-306 (especially 304-305), Cosío to Terán (6-Jun-1830) frames 371-376, Elosúa to Cosío (8-Jun-1830) frames 411-412, Garcia to Cosío (14-Jun-1830) frames 649-651, Galan to Cosío (17-Jun-1830) frames 723-724, Cosío to Elosúa (18-Jun-1830) frames 743-756 (especially 747-750), Elosúa to Cosío (21-Jun-1830) frames 823-836 (especially 827-828), Cosío to Elosúa (29-Jun-1830) Roll 132 frames 37-41, and Terán to Elosúa (15-Jul-1830) frames 578-596 (especially 581-582)

De Langara, Juan,
1799   Map, Carta Esferica que comprehende las costas del Seno Mexicano Construida De Orden Del Rey En El Deposito Hydrografico De Marina; image of 1805 version available on-line at https://www.raremaps.com/gallery/detail/64047/carta-esferica-que-comprehende-las-costas-del-seno-mexicano-direccion-hidrografica-de-madrid?q=0, accessed 21-Oct-2022

De Las Piedras, José,
2-Aug-1830   Letter to Elosúa, Antonio, **BA**, Microfilm Roll 133, frames 58-65 (especially 61-65); see also Elosúa to Cosío & Arteaga (17-Aug-1830) frames 596-607 (especially 6060), Piedras to Elosúa (13-Sep-1830) Roll 134 frames 394-406 (especially 396-399) & Elosúa to Chowell (29-Sep-1830) frame 844

Del Valle, Santiago,
7-May-1832   Letter to Ramón Músquiz, **BA**, Microfilm Roll 149, frames 826-831 (especially 830-831)

Elosúa, Antonio,
2-Mar-1830   Letters to Manuel de Mier y Terán & Mariano Cosío, **BA**, Microfilm Roll 128, frames 879-880 & 881-882
9-Jun-1830   Letter to Martin de Leon, and reply (18-Jun-1830), **BA**, Microfilm Roll 131, frames 442 and 742
21-Jun-1830   Letter to Mariano Cosío, **BA**, Microfilm Roll 131, frames 823-836
28-Jun-1830   Letter to Manuel de Mier y Terán, **BA**, Microfilm Roll 132, frame 13
5-Jul-1830   Letter to Manuel de Mier y Terán, **BA**, Microfilm Roll 132, frames 301-306
21-Jul-1830   Letter to Aniceto Arteaga, **BA**, Microfilm Roll 132, frames 768-773 (especially 770)
19-Aug-1830   Letter to Aniceto Arteaga, **BA**, Microfilm Roll 133, frames 686-687
21-Oct-1831   Letter to Aniceto Arteaga, **BA**, Microfilm Roll 145, frames 478-481, see also Arteaga to Elosúa (31-Oct-1831) & Cosío to Elosúa (4-Nov-1831), frames 675-681 & 757-762

27-Mar-1832 Letter to Manuel de Mier y Terán, **BA**, Microfilm Roll 148, frame 959; see also Cosío to Elosúa (6-Apr-1832) Roll 149, frames 184-193 (especially 184-185)

16-Apr-1832a Letters to Ramón Músquiz & Erasmo Seguin, **BA**, Microfilm Roll 149, frames 486-493; see also Músquiz to Commandants at Goliad & Lavaca (16-Apr-1832) Roll 149 frames 498-499, Músquiz to Manchola & De Leon (17-Apr-1832) Roll 149, frames 509-510, De Leon to Arteaga (3-May-1832) Roll 149 frames 745-746, Arteaga to Elosúa (4-May-1832) Roll 149 frames 772-773

16-Apr-1832b Letter to Aniceto Arteaga, **BA**, Microfilm 149, frames 494-497

30-Jun-1832 Letter to Aniceto Arteaga, **BA**, Microfilm 151, frames 189-199 (especially 194)

16-Aug-1832 Letter to Ignacio de Mora, **BA**, Microfilm Roll 152, frames 628-629

Fernandez Rincón, Augustin,
24-Apr-1832 Letter to Manuel de Mier y Terán, **BA**, Microfilm 149, frames 622-623

Filisola, Vicente,
1848 Memorias para la historia de la guerra de Tejas, Mexico City, 2 volumes, 1848; English translation is Memoirs For The History Of The War In Texas, translated by Wallace Woolsey, 2 volumes (Eakin Press, Austin, Texas), 1985, especially Vol. I, pp. 62-89, & 139

Fisher, George,
5-Jun-1830 Letter to Samuel May Williams, Photostat of handwritten copy by George Fisher on 6-Nov-1830, found in **GFP** as "Port of Galveston report", Box 2.325/V36; exact same "original" document is found in **SMWC**, Manuscript 23-0466, but last page in latter is the same as first page of former.

Guerra, José Mariano,
10-Jun-1831 Letter to Antonio Elosúa, **BA**, Microfilm Roll 141, frames 832-834, see also Guerra to Elosúa (9- and 13-Jun-1831), Roll 141, frames 805-806 and Roll 131, frames 603-604

20-Jun-1832 Letter to Antonio Elosúa and Mariano Cosío, **BA**, Microfilm Roll 150, frames 927-930 and 931; see also Guerra to Elosúa (27-Jun-1832) Roll 151, frames 73-74 and 77

27-Jun-1832 Letter to Antonio Elosúa, 27-Jun-1832, Béxar Archives, Microfilm Roll 151, frames 73-76; see also Elosúa to Anahuac/Lavaca/Goliad and to José de las Piedras (5-Jul-1832) Roll 151, frame 373 and 378

Hernandez, Juan José,
10-Aug-1832 Letter to Ramón Músquiz, **TGLO**, Document# 1190, Box 128/6, p. 108

Howren, Alleine,
Apr-1913 "Causes and Origin of the Decree of April 6, 1830", **SWHQ**, Vol. XVI (16), No. 4, pp. 378-422

Jackson, Jack, Robert S. Weddle, Winston De Ville,
1990 *Mapping Texas And the Gulf Coast, The Contributions of Saint-Denis, Oliván, and Le Maire*, (Texas A&M University Press, College Station TX)

Jackson, Jack, Margaret Howard, Luis A. Alvarado,
Apr-2006 *History and Archeology of Lipantitlán State Historic Site, Nueces County, Texas*, Texas Parks and Wildlife Dept. Cultural Resources Program, Apr 2006

Lewis, W. S.,
Jul--1899 "The Adventures of the 'Lively' Immigrants", **SWHQ**, Vol. III (3), No.1 (Jul 1899), pp. 1-32
Oct-1899 and No.2 (Oct 1899), p. 81-107

Linn, John J.,
1883 *Reminiscences of Fifty Years In Texas*, State House Press, 1986, copy of the original 1883 edition

Manchola, Rafael,
29-Jul-1826 Letters to Mateo Ahumada, **BA**, Microfilm Roll 95, frames 478-481; also see Ahumada to Austin (3-Aug-1826) & reply (25-Aug-1826), **AP**, Vol. I, pp. 1398 & 1439.

Martin, Robert Sidney,
Apr-1982a "Maps of an Empresario: Austin's Contribution to the Cartography of Texas", **SWHQ**, Vol. LXXXV (85), No. 4 (Apr 1982a), pp. 371-400

Martin, Robert Sidney, James C. Martin,
1982b *Contours of Discovery – Printed Maps Delineating the Texas and Southwestern Chapters in the Cartographic History of North America, 1513-1930 - A User's Guide* (1982b, Texas State Historical Assoc1ation & Center for Studies in Texas History at UT Austin)

Menefee, John S.,
1880 "Early Jackson County History", typescript manuscript for articles in ***Jackson County Clarion*** newspaper (Texana, Texas), 20-May to 15-Jul-1880, William Ransom Hogan Papers, Box GA21, Folder 11, University of Texas at Arlington, Special Collections Library; a copy is also available in the John S. Menefee Papers, Box 2R116 at the Dolph Briscoe Center for American History

Mexía, José Antonio,
18-Jul-1832 Letter to Antonio Elosúa, **BA**, Microfilm Roll 151, frame 700; see also Paredes to Elosúa (13-Jul-1832) frames 583-584, Guerra to Elosúa (23-Jul-1832) frames 810-814 (especially 810-812), Elosúa to Cosío & Piedras (24-Jul-1832) frames 852-5, Elosúa to Mexía (2-Aug-1832) Roll 152 frame 243

Midkiff, George,
1-Jul-1830 Letter to Mariano Cosío, **BA**, Microfilm Roll 132, frames 113-114; see also Cosío to Elosúa (2-Jul-1830) frames 136-155 (especially 145-146), Elosúa to Terán (6-Jul-1830) frames 363-375 (especially 368-369), Elosúa to Cosío (7-Jul-1830) frames 442-448 (especially 447), Cosío to Terán (16-Jul-1830) frames 647-654 (especially 647-648), Terán to Elosúa (21-Aug-1830) frames 710-717 (especially 710-711), Terán to Seguin (21-Aug-1830) frames 718-721 (especially 718-719)

Moret, Juan,
6-Aug-1832   Letter to Mariano Cosío, **BA**, Microfilm Roll 152, frame 390; see also Cosío to Elosúa (7-Aug-1832) frames 398-401, Elosúa to de Mora (9-Aug-1832) frame 453 and Elosúa to Cosío (9-Aug-1832) frames 454-455

Morton, Ohland,
1944, 1945   "Life of General Don Manuel de Mier y Terán, as it affected Texas-Mexican Relations, Chapter IV – The Law of April 6, 1830 and Chapter V --- Affairs in Texas, 1831-1832", **SWHQ**, Vol. XLVII (47), No. 2 (Apr 1944), pp. 193-218 and Vol. XLVIII (48), No. 4 (Apr 1945), pp. 499-546; these chapters (and the others) published in book form as *Terán and Texas* (Texas State Historical Association, Austin, Texas, 1948)

Músquiz, Ramón,
19-Jul-1830   Letter to José Maria Viesca, **BA**, Microfilm Roll 132, frame 710
24-May-1832   Letters to Alcaldes of Goliad & Austin, **BA**, Microfilm Roll 150, frames 218-221 (especially 218-221); see also Campos to Flores & Músquiz (7-May-1832) Roll 149, frames 826-831
30-Jun-1832   Letters to Aniceto Arteaga, 30-Jun & 2-Jul-1832, English translation, **RBBRC**, Supplement Vol. XII, pp. 384-385 and 400-401; 30-Jun-1832 original found in **BA**, Microfilm Roll 151, frame 202

Quintanilla, Romualdo,
24-Jun-1832   Letter to Aniceto Arteaga, English translation, **RBBRC**, Vol. XII, pp. 352

Reinhartz, Dennis,
2015   "Maps of Stephen F. Austin: An Illustrated Essay of the Early Cartography of Texas", *The Occasional Papers*, Series No. 8 (Winter 2015), a Philip Lee Phillips Map Society Publication, Library of Congress – Geography and Maps Division

Ruiz, José Francisco,
26-Dec-1830   Letter to Samuel May Williams, English translation in **PCRCT**, Vol. 5, 1978, pp. 352-353; Spanish original is found in **SMWC**, Manuscript 23-0503

Seguin, Erasmo,
16-Apr-1832   Letter to Antonio Elosúa, **BA,** Microfilm Roll 149, frames 480-481

Shook, Robert W.,
2007   *Caminos y Entradas – Spanish Legacy of Victoria County and the Coastal Bend 1689-1890*, Victoria County Heritage Department

Soler, Miguel Cayetano,
28-Sep-1805   Letter transmitting King of Spain's authorization for free trade at Bahia de San Bernardo, **BA**, original in Box 2S81; microfilm copy on Roll 33, frames 652-653; CAH translation in Box 2C6, Series II, General Manuscripts, Vol. VIII, p. 420; microfilm of translation on Reel 20; also see Salcedo to Cordero (11-Feb-1806) Roll 34 frame 282 & Iturrigaray (25-Feb-1806) Roll 34 frames 345-346

Spanish Admiralty,
1807   Map, Carta particular de las Cóstas Setentrionales del Seno Mexicano que comprehende las de la Florída Ocidental las Margenes de la Luisiana y toda la rivera que sigue por la Bahia de S. Bernardo y el Rio Bravo del Norte hasta la Laguna Madre, Madrid; copy found as Map# 197 at Rosenberg Library (Galveston TX) & digital scans made from it

Taliaferro, Henry G. (compiler),
1988   *Cartographic Sources in the Rosenberg Library*, edited by Jane A. Kenamore & Uil Haller, (Texas A&M University Press, College Station TX), especially Map #201, p. 99; Rosenberg Library has 93 of the 121 maps of the Portulano de la America Setentrional, including #33 Bahia de S. Bernardo (modern Matagorda & Lavaca Bays) and #34 Bahia de Galveztowm (modern Galveston Bay, Rosenberg Map# 200)

Taylor, Ira Thomas,
1938   *The Cavalcade of Jackson County*, The Naylor Co., San Antonio, Texas

Terán, Manuel de Mier y,
25-Feb-1830   Letter to Antonio Elosúa, **BA**, Microfilm Roll 128, frames 686-690 (especially frames 688-690); see also Elosúa to Terán (11-Mar-1830) Roll 129, frames 20-29 (especially 20-21) & Elosúa to Cosío (11-Mar-1830) frames 30-33 (especially 32)
24-Apr-1830   Letter to Antonio Elosúa, **BA**, Microfilm Roll 129, frames 965-966; see also Elosúa to Cosío (27-Apr-1830) Roll 130 frames 65-72 (especially 67), Elosúa to Terán (4-May-1830) frames 200-203 (especially 202-203), Elosúa to Cosío (4-May-1830) frames 204-211 (especially 208), & Terán to Elosúa (3-Jul-1830) Roll 132, frames 180-183 (especially 180-181)
25-May-1830   Letter to Antonio Elosúa, **BA**, Microfilm Roll 130, frames 672-675
31-May-1830   Letters to Andres Sobrevilla (pp.64-65), Martin de Leon (p. 66) and Lucas Alamán (p. 67), **AGMC**, Box 2Q171, Volume 327; also see Letter from Terán to Cosío, insertion in Cosío to Elosúa, 2-Jul-1830, **BA**, Microfilm Roll 132, frames 143-144b; and Elosúa to Terán, 6-Jul-1830, Microfilm Roll 132, frames 371-372
1-Jun-1830   Letters to Antonio Elosúa (frames 838-841) and Erasmo Seguin (frames 842-849), **BA**, Microfilm Roll 130, Dolph Briscoe Center for American History, The University of Texas at Austin, Austin TX; instructions to Rafael Chowell also found (pp. 4-5) and Letter to Stephen F. Austin (pp. 7-8), 1-Jun-1830, **AGMC**, Box 2Q223, Volume 562; another copy of instructions to Chowell found in **TGLO**, Document# 598, Box 125/10, p. 112
5-Jun-1830   Letter to Antonio Elosúa, **BA**, Microfilm Roll 131, frames 332-333 and 334-335
17-Jul-1830   Letter to Antonio Elosúa, **BA**, Microfilm Roll 132, frames 664-667
21-Aug-1830   Letter to Miguel Muro, insertion in letter to Antonio Elosúa, **BA**, Microfilm Roll 133, frames 716-717b
15-Sep-1830   Letter to Rafael Chowell, enclosure in Letters to Elosúa and Seguin, **BA**, Microfilm Roll 134, frames 440-444 and 445-446
19-Sep-1830   Letter to Aniceto Arteaga, enclosure in Letters to Chowell, Seguin, Elosúa and Seguin, **BA**, Microfilm Roll 134, frames 514-519
20-Sep-1830   Letter to Antonio Elosúa, English translation in **PCRCT**, Vol. IV, 1977, pp. 486-487; original found as **TGLO**, Document# 614, Box 125/11, p. 131; see also Ruiz to Elosúa (Document# 615, 9-Oct-1830, 125/11 p. 132) and Elosúa to Terán (Document# 604, 9-Dec-1830, 125/10 p.119)

12-Oct-1830	Letter to Antonio Elosúa, English translation in **PCRCT**, Vol. V, 1978, pp. 68-69; original found as **TGLO**, Document# 601, Box 125/11, p. 116
30-Oct-1830	Letter to José Manuel Barberena, enclosure in letter from Barberena to Elosúa, 15-Nov-1830, **BA**, Microfilm Roll 136, frames 247-248
30-Dec-1830	Letter to José María Viesca, English translation in **PCRCT**, Vol. V, 1978, pp. 360-362; see also Viesca to Músquiz, 22-Jan-1831, **TGLO**, Document# 586, Box 125/9
15-May-1831	Letter to Ramón Músquiz, **TGLO**, Document# 596, Box 125/9, p. 110
16-May-1831	Letter to Antonio Elosúa, **BA**, Microfilm Roll 141, frames 216-217; see also Guerra to Seguin and Elosúa (16-Jun-1831), Roll 141, frames 973-974\
24-Sep-1831	Letter to Antonio Elosúa, **BA**, Microfilm Roll 144, frames 717-718; see also Barberena to Elosúa, (20-Oct-1831), Elosúa to Terán (21-Oct-1831), Elosúa to Barberena (21-Oct-1831), Elosúa to Barberena (24-Oct-1831) & Elosúa to Barberena (15-Nov-1831), Roll 145, frames 443-444, 470-477, 486-487, 547-548 & 978
12-Apr-1832	Letter to Antonio Elosúa, **BA**, Microfilm Roll 149, frame 348
2000	*Texas by Terán – The Diary Kept by General Manuel Mier y Terán on his 1828 Inspection of Texas* (English translation, edited by Jack Jackson, University Of Texas Press, 2000)

Trujillo, Miguel Zenon,

3-Sep-1830	Commissary Review Report for convicts at Guadalupe, **BA**, Microfilm Roll 134, frames 91-95
3-Jan-1831	Commissary Review Report for convicts at Barranco Colorado, **BA**, Microfilm Roll 137, frames 588-590

Turner, F. H.,

Jul-1903	*The Mejía Expedition*, **SWHQ**, Vol VII (7), No. 1, pp. 1-28

Ugartechea, Domingo de,

8-Jul-1832	Letter to Aniceto Arteaga, forwarded to Elosúa on 10-Jun-1832, **BA**, Microfilm Roll 151, frames 518-519

Weddle, Robert S.,

1992	*Spanish Exploration of the Texas Coast, 1519-1800*, Bulletin of the Texas Archeological Society, Vol. 63 (1992), pp. 99-118
1999	*Wilderness Manhunt – The Spanish Search for La Salle*, (1999, Texas A&M University Press, College Station, Texas)

White, Joseph M.,

1839	*A New Collection Of Laws, Charters and Local Ordinances Of The Governments Of Great Britain, France and Spain, Relating To The Concessions Of Land In Their Respective Colonies; Together With The Laws Of Mexico And Texas On The Same Subject* (T. & J. W. Johnson, Philadelphia PA), especially Vol. I, Book V, pp. 421-622

INDEX

Adams-Onis Treaty of 1819, 2
Alamán, Lucas, 2, 26, 59
Alligator Head, 15
Anahuac, 26, 42, 44, 45
Anahuac Disturbances, 42, 44
Aransas
    Bay, 9, 13, 14, 46
    Pass, 24
    Point, 33
    town of, 46
Aransazu Bay, 14
Aranzas Bay, viii, 14
Aranzazu, 28
Arenosa Creek, 23
Arteaga, Aniceto, vi, viii, 20, 21, 22, 23, 24, 25, 26, 27, 28, 29, 30, 32, 33, 35, 38, 39, 40, 41, 42, 43, 44, 45, 49, 51, 52, 53, 55, 56, 58, 59, 60
Arze, Juan Perez de, 36
Asqueta, Luis, 25
Atascosito Road, 46
Austin Papers, v, 50
Austin, Moses, 9
Austin, Stephen F., ii, viii, 1, 2, 3, 8, 9, 10, 11, 12, 13, 14, 26, 27, 28, 42, 46, 50, 51, 52, 53, 56, 57, 58, 59
Austin's Colony, 1, 9, 25, 26
Bahía de San Jose, 9
Bailey, James B. "Britt", 42
Balandra Point, 12
Barberena, José Manuel, viii, 26, 27, 33, 39, 40, 49, 53, 60
Barranco Colorado, iii, v, vi, viii, 1, 3, 4, 19, 22, 23, 24, 26, 36, 38, 39, 41, 45, 46, 47, 48, 49, 50, 53, 60
Barroto, Juan Enríquez, 4
Bayucos Island, 15
Becerra, José María, 25
Berlandier, Jean Louis, i, viii, 2, 12, 13, 14, 45, 46, 50, 53
Béxar Archives, v, vi, 19, 23, 27, 29, 30, 36, 39, 40, 41, 43, 44, 50, 56
Boujean, Pierre, 24
Brazoria
    County Historical Museum, v, 54
    County Library System, v
    town of, v, 1, 20, 42, 43, 44, 54

Brazos
    River, vi, 3, 6, 8, 9, 10, 12, 13, 28, 41, 42, 44
Brazos Santiago, 24, 25
brick, 20, 21, 22, 23, 26, 38, 44, 47, 48, 49
Brick Factory Springs, 47
Cabeza de Vaca, 4
Calhoun, town of, 46
***Cañon*, schooner**, 28, 41, 43, 45
Carbajal, José M. J., 42
Carlos Bay, 13
Castillo, Jóse María, viii, 28, 29, 30, 31, 36, 39, 41, 53
cemetery, 43, 49
Chowell, Rafael, i, viii, 20, 21, 26, 27, 28, 30, 32, 33, 34, 36, 38, 41, 43, 44, 49, 51, 53, 55, 59
Coahuila y Tejas, ii, 3, 12, 13
Colorado River, vi, 4, 5, 6, 8, 9, 12, 13, 16, 18, 19, 20, 23, 38, 39, 41, 46, 47, 54
Comandancia Militar del Establecimiento de la Vaca, 24, 49, 50
Comisión de Límites, i, 2, 14
***Constante*, schooner**, 24, 26, 27, 30, 33, 35
Copano Bay, 9, 13, 14, 46
Cópano, port of, 13, 14, 24, 41, 45, 46
Corpus Christi, 25
Cosío, Mariano, 24, 25, 26, 27, 28, 40, 41, 42, 43, 51, 54, 55, 56, 57, 58, 59
Coushatta Trace, 2, 46
Cox's Point, 19, 20, 46
Culebras, Punta de, 4, 5, 9, 13
Dimitt's Landing, 22
Dimmit's Landing, 46
Dog Island Reef, 16
Dolph Briscoe Center for American History, v, 8, 10, 50, 57, 59
East Matagorda Bay, 19
Eastern Interior Provinces of Mexico, 2, 8, 12
Edna, Texas, 47
El Camino Real, 11
Elosúa, Antonio, 24, 25, 26, 27, 28, 29, 38, 39, 40, 41, 42, 44, 45, 51, 52, 53, 54, 55, 56, 57, 58, 59, 60
Espiritu Santo Bay, 3, 4, 13, 14, 46
Evia, José de, viii, 5, 6, 55
Filisola, Vicente, 3, 4, 56
Fisher, George, 12, 28, 41, 50, 56
Flores, Salvador, 24, 28

Fort Anahuac, 3
Fort Lavaca, 4
Fort Lipantitlán, 3, 4, 38
Fort St. Louis, 4, 6
Fort Tenoxtitlán, 3, 26, 38
Fort Terán, 3
Galan, José Bonifacio, 25, 39, 41, 43
Galveston
    Bay, 3, 6, 9, 12, 26, 44, 59
    Island, 3, 6, 9, 12
    port of, 13
Galvezton, Aduana Maritima de, 41
Galvezton, el puerto de, 12
Garcia, José Valentin, 25
Garcitas anchorage, 25, 35
Garcitas Cove, 15
Garcitas Creek, 4, 6, 19, 35, 42
Garza, Cayetano, 42
Garza, Francisco de la, 25
Garza, José María de la, 24
General Bustamante, schooner, 26, 27, 29
Goliad, i, viii, 8, 25, 26, 27, 33, 36, 39, 40, 41, 42, 43, 44, 45, 53, 56, 58
Gonzales, town of, 2, 52
Green DeWitt Colony, 23
Guadalupe River, 26, 49
Guadalupe Victoria, viii, 2, 22, 23, 24, 27, 28, 29, 30, 33, 36, 38, 39, 40, 41, 42, 43, 44, 46, 49
Guerra, José Mariano, 39, 42, 43
Gulf of Mexico, viii, 5, 7
Harrisburg, town of, 1, 4, 46
Hatch, Sylvanus, 19
Hernandez, Juan José, 39
Hetta (aka Hesta), schooner, viii, 19, 25, 26, 39
Hetta (aka Hesta). schooner, 40
Humboldt map, 9, 10
Hunt-Randel map of 1839, viii, 14, 46
Indianola, 5, 12
Jackson County, iii, v, viii, 19, 20, 21, 22, 23, 47, 50, 52, 57, 59
Jaraname tribe, 38
Karankawa tribe, 38
Keeran Point, 15
La Bahía, 13, 26, 27, 29, 33, 45, 49
La Belle, 4, 5
Lamar, town of, 46
Lampazos, 26
Langara, Juan de, viii, 6, 7, 55
Laredo, 2, 26
Lavaca
    anchorage, 26, 27
    Bay, viii, 3, 4, 5, 9, 13, 14, 15, 16, 19, 25, 35, 39, 41, 44, 46
    detachment or post or establishment, 19, 21, 23, 26, 28, 29, 33, 36, 37, 38, 39, 42, 43, 45
    Port, 4, 5, 19
    River, iii, 1, 3, 4, 5, 6, 10, 11, 15, 19, 20, 22, 23, 24, 26, 27, 46, 47, 48, 49
Law of 6-Apr-1830, 2, 27, 49
León, Alonso de, 4, 5
Leon, Fernando de, 25
Leon, Martin, 26, 27, 49, 55, 59
Leona Victoria, Mexico, 41
Lewis, Nathaniel, 19, 25
Library of Congress, iii, 9, 10, 58
Linn, Charles, 24
Linn, Edward, 23
Linn, John J., vi, viii, 20, 21, 22, 23, 24, 39, 41, 44, 57
Linnville, town of, 5, 19, 46
Lively, 12
Lopez, Miguel, 42
Los Mosquitos, 25, 26
Mad Island Reef, 16
Manchola, Rafael, 39
Manso, José Dominguez, 10
Manso, Leonardo, 23
Matagorda
    Bay, viii, 3, 4, 5, 6, 8, 9, 12, 14, 15, 16, 17, 18, 19, 25, 44, 46, 52
    detachment of, 12
    Peninsula, 16, 19
    Port of, 24, 25, 26, 28, 41, 49
    Punta de, 13, 25
    town of, 16, 19
Matamoros, i, 2, 3, 21, 24, 25, 26, 27, 28, 36, 39, 40, 41, 42, 44, 46
McCallick, Mindora Bagby, 21
Menefee, John S., 19, 39
Mexía, José Antonio, 21, 44
Mexicana, schooner, 12, 24
Mexico, i, vi, viii, 3, 8, 21, 22, 43, 45, 50, 53, 55, 56, 60
Mexico City, 12
Midkiff, George, 24, 25, 57
Mier, 26, 42
Molly, John, 41
Moret, Juan, 44, 45
Muro, Miguel, 29, 38, 59

Músquiz, Ramón, 3, 25, 26, 41, 42, 43, 52, 54, 55, 56, 58, 60
Nacogdoches, 2, 3, 4, 8, 9, 11, 28, 36, 42, 45, 52
Navarro, Eugenio, 25
Navarro, José Antonio, 13
Navarro, José Eugenio, viii, 14, 24, 25
Navayra, Santiago, 38
Navidad River, 19, 46
New Orleans, 12, 13, 14, 19, 24, 25, 28, 45
Nieto, Miguel, 44
Old Station Road, 23
Old Three Hundred, 1, 12
Oposición, schooner, 24, 25, 55
Oscar, schooner, 24, 26, 28
Osores, Manuel, 33
Padilla, Vicente, 42
Paso Cavallo, 4, 15, 19, 41, 49
Pettit, Edward, 41
Piedras, José de las, 36
Pike, Zebulon map, 9, 10
Pomona, schooner, 13, 14, 28
Port O'Connor, 5, 12, 15
Portal to Texas History, v, 8, 17
Presidio La Bahía, 3
Puelles, Fray José Maria, viii, 8, 9, 10
Quintanilla, Romualdo, 42
Red Bluff, town of, 47
Rover, schooner, 28
Ruiz, José Francisco, 38
Sabine River, 12
Sabinito, 25, 26
Sabino, 12
Saltillo, 12, 41
San Antonio Bay, 9, 12, 25, 26, 46
San Antonio de Béxar, 2, 3, 11, 13, 25, 26, 45
San Antonio River, 8, 26
San Bernard River, 42
San Bernardo
   Bahía de, viii, 4, 5, 8, 9, 12, 13, 17, 19, 27, 49
   Lago de, 5, 6
San Felipe de Austin, 2, 4, 22, 28, 38, 42, 43, 46
San Francisco, Punta de, 4, 5, 13
San Jacinto River, 11
San Luis Pass, 6
San Luis, Isla de, 5, 6, 8, 9
San Marcos, Rio de, 5
Santa Anna, Antonio López de, vi, 21, 43, 44
Second Flying Company of San Carlos de Alamo de Parras, 14
Seguin Erasmo, 26

Seguin, Erasmo, 26, 27, 33, 41, 43
Shell Island Reef, 16, 17
Shook, Robert W., v, 4, 5, 10, 23, 44, 46, 58
Sobrevilla, Andres, 26, 59
Sol, schooner, 24
Spanish Admiralty, viii, 8, 59
Subarán, Felix, 44
Tamaulipas, 43
Tamaulipas, 3rd Active Company of, 26, 33
Tampico, 2, 3, 26, 30, 39, 44
Taylor, Ira Thomas, 22
Tenoxtitlán, 26, 28, 38
Terán, Manuel de Mier y, i, ii, vi, viii, 1, 2, 3, 4, 12, 14, 20, 24, 25, 26, 27, 28, 29, 34, 36, 38, 39, 40, 41, 42, 43, 44, 45, 46, 47, 49, 51, 54, 55, 56, 57, 58, 59, 60
Texana Road, 22, 23
Texana, town of, 19, 22, 46, 47
Texas, i, ii, vi, viii, 1, 2, 3, 4, 5, 6, 7, 8, 9, 12, 14, 15, 16, 17, 20, 21, 23, 24, 25, 26, 27, 28, 36, 38, 39, 41, 43, 44, 45, 46, 47, 49, 50, 51, 52, 53, 54, 56, 57, 58, 59, 60
Texas Archeological Research Laboratory, 47
Texas Gazette, viii, 2, 28, 47, 52, 54
Texas General Land Office, v, 14
Texas Republican newspaper, The, viii, 20
Texas Revolution, vi, 20, 22, 45, 49
Thompson, Thomas M., 25
tobacco, 25, 28, 41
Trinity Bay, 9
Trinity River, 3, 46
Trujillo, Miguel Zenon, viii, 33, 34, 36, 60
U.S. Coastal Survey, 15
Ugartechea, Domingo de, 43, 44
Velasco
   Battle of, 21, 42, 43, 44
   Fort, vi, 3, 4, 42, 45, 49
   Fortaleza de, vi
   town of, vi, 42, 43, 44, 45
Venado Creek, 15
Veracruz, 3, 4
Victoria
   County, viii, 22, 23, 58
   town of, 10, 20, 24, 27, 46
Viesca, José María, 3, 25, 53, 55, 60
Villasana, Lt. Col., 21
Williams, Samuel May, 38
Zacatecas, 26
Zorra, schooner, 24, 25

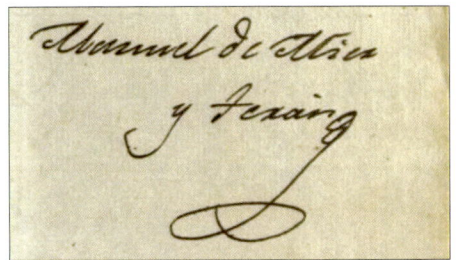

Manuel de Mier y Terán's signature, Béxar Archives 130:673

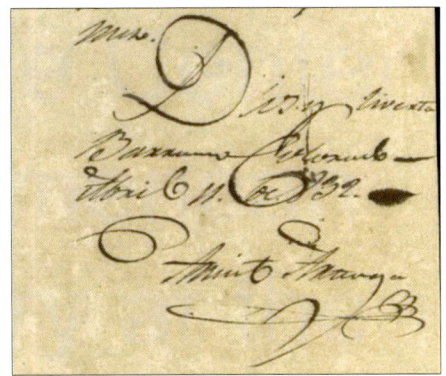

Dios y Libertad salutation and Aniceto Arteaga's signature, Béxar Archives 149:335

Rafael Chowell's signature, Béxar Archives 133-674

José Manuel Barberena's signature, Béxar Archives 140-716

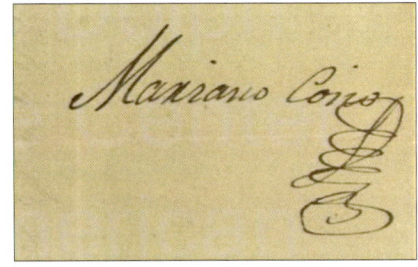

Mariano Cosío's signature, Béxar Archives 150:310